LODE GOLD MINES OF THE ALLEGHANY-DOWNIEVILLE AREA, SIERRA COUNTY, CALIFORNIA

By Denton W. Carlson * and William B. Clark *

OUTLINE OF REPORT

Illustrations

ABSTRACT

The Alleghany-Downieville area is one of the few remaining areas in California where lode gold mining is a major source of income. In 1953, more than $700,000 in gold was produced from the lode mines of the area. The total lode gold production from the Alleghany-Downieville region has been in excess of $50,000,000. Lode mines which have had the greatest amount of activity and production since World War II are the Sixteen-to-One, Brush Creek, Kate Hardy, Oriental, and Red Star mines.

The area, approximately 125 square miles in extent, lies on the western slope of the Sierra Nevada in southwestern Sierra County. It is characterized by rugged topography, and elevations range from 2,500 to 6,000 feet.

---

* Assistant Mining Geologist, California Division of Mines.

Bedrock in the area consists of steeply dipping, isoclinally folded metamorphosed volcanic and sedimentary rocks of Carboniferous and Permian age. Intrusive into these are Jurassic igneous rocks, chiefly ultrabasic rocks which have since altered to serpentine. Overlying the bedrock are gently west-dipping Tertiary and Quaternary volcanic and sedimentary rocks.

The Alleghany veins occur as quartz-filled fissures and replacement bodies in a fracture system that was formed principally by north-northwest-trending reverse faults. These veins dip east or west with the east-dipping veins tending to be flatter and more persistent. The Downieville veins dip steeply and contain larger amounts of sulfide minerals than the Alleghany veins.

The most characteristic features of the ore deposits in the Alleghany district are the extreme richness, very erratic distribution, and small size of the ore shoots. Most of the gold produced from the mines in the Downieville district has been from sulfide-rich ores, and a higher proportion of the gold occurs in relatively large but lower grade ore shoots than those in Alleghany. Certain favorable structural features control the ore deposition in the Alleghany veins. These include irregularities in the attitude of the vein and minor faulting. Serpentine commonly has had a direct influence in the location of highgrade ore.

The mineralization in the Alleghany district has been divided into four stages: Chlorite, quartz, carbonate, and final stages. During the carbonate stage, there was extensive replacement of the wall rocks by carbonate as well as deposition of the major portion of the gold. Economically the carbonate stage was the most important of all the stages.

In mining the Alleghany veins, the more favorable portions of the vein are cut up into small blocks by means of drifts and raises and the ore thus found is stoped. All of the mines are developed by adits. Little support timbering is required in the underground workings except in serpentine.

Gold ore is milled by crushing in jaw crushers, grinding in ball mills, gravity concentration, amalgamation, and flotation. Mill-head assays range from less than $5 to $30 or more in gold per ton. Mill recoveries range from about 80 to 96 percent. Most of the high-grade ore is milled separately.

INTRODUCTION

The Alleghany-Downieville district of Sierra County is one of the few remaining areas in California where lode gold is a major source of income. During 1953, the lode mines in this area produced more than $700,000 in gold. Since World War II, approximately 15 lode mines have been active at least part of the time. At present, the largest gold producers are the Sixteen-to-One and Brush Creek mines.

Gold has been produced from this area since the earliest days of California gold mining. Since 1907, the chief source of gold has been from the lode mines (MacBoyle, 1918, p. 1). Most of the lode production has been from small but rich ore shoots in the veins; relatively little gold has been obtained from the veins outside of the ore shoots. The total lode gold production from the Alleghany-Downieville area has been in excess of $50,000,000, most of which is based on a price of $35 per ounce.

Geography. The Alleghany-Downieville area is in southwestern Sierra County about 25 airline miles northwest of Grass Valley. Downieville, the county seat of Sierra County, and Alleghany, several airline miles to the south, are the principal towns in the area. Other towns include Goodyears Bar and Forest City.

State Highway 49 crosses the area from east to west and connects Downieville with Grass Valley, 50 miles to the southwest. Alleghany is connected to Goodyears Bar, which is 4 miles west of Downieville, by a 10-mile unpaved road. Alleghany is also served by the Ridge road, which extends 18 miles southwest along Pliocene Ridge to join State Highway 49, and the Foote road which runs west along the Middle Fork of the Yuba River and then south to North Columbia.

The region is drained by the southwest-flowing North and Middle Forks of the Yuba River and numerous subsidiary streams. All have cut deep canyons with steep walls. Elevations above sea level range from 2,500 feet in the canyons to more than 6,000 feet on the crests of the ridges. Downieville and Alleghany are at an elevation of 3,000 and 4,400 feet, respectively.

History. Gold has been produced in the Alleghany-Downieville area since 1849. Soon after the beginning of the California gold rush, rich discoveries along the many streams attracted thousands of gold seekers to the area. Among them was William Downie who established a settlement known now as Downieville. Sierra County was established in 1852. The river bars were mined first, after which hydraulicking and drift mining of the buried auriferous channels were done. The peak of the early production at the Alleghany area was about 1861 (Ferguson and Gannett, 1932, p. 25). Subsequently, the more deeply buried gravel deposits of the Forest district were exploited.

Lode gold mining began in the area in 1853 (Ferguson and Gannett, 1932, p. 26), and during the 1870's, such lode mines as the Plumbago and Brush Creek produced substantial amounts of gold. The rediscovery of the Tightner vein in 1907 caused a revival of lode gold mining in the area, which is now continuing. The greatest year of lode gold production for the district was in 1922 when $1,710,521 were produced (Ferguson and Gannett, 1932, p. 27). In the late 1930's, drift mining was revived with such mines as the Ruby contributing substantially to the total gold production of the area.

In 1942, War Production Board Order L-208 was placed in effect and all of the gold mines either shut down or remained open under limited operation. Since World War II, some of the mines have resumed operations on a scale as great as that prior to the war. The

FIGURE 1.   Road cut exposing Tightner formation
near Red Star mine.

post-war increase in wages and material costs coupled with a fixed price of gold and the expense of reconditioning idle mines have caused a number of the mines to remain idle. Former important lode mines now idle include the Rainbow and Plumbago.

GEOLOGY

The oldest and most extensive rocks of the region are a series of metamorphosed sedimentary and volcanic rocks of the Calaveras group (Carboniferous, in part). They have a north to northwest trend, dip steeply east, are isoclinally folded and in part overturned to the west. In this area the Calaveras group has been divided into five formations of which three, the Blue Canyon, Relief, and Cape Horn, were mapped and described by Turner (1897) and Lindgren (1900) and two, the Tightner and Kanaka, were mapped and described by Ferguson and Gannett (1932, pp. 6-12).

Intrusive into the Calaveras group are ultrabasic and basic igneous rocks and small amounts of granitic rocks. Virtually all of the ultra-basic igneous rocks have been altered to serpentine.

Overlying the Calaveras group and granitic and basic igneous rocks, especially on the interstream ridges, are nearly flat-lying auriferous gravels (Eocene to Miocene), andesitic breccia (Mio-Pliocene), basalt flows (Pleistocene), and minor amounts of stream gravels (Pleistocene to Recent).

Rock Formations
Calaveras Group (Carboniferous in Part)

Blue Canyon Formation. The Blue Canyon formation occurs extensively in beds of great thickness east of the Alleghany-Downieville area and is the basal formation of the Calaveras group in the area.

The formation is composed chiefly of dark slate which becomes nearly white on weathering. In places, the slate grades into micaceous schist. Present in small amounts are fine-grained quartzite, impure limestone, and fine-grained conglomerate. The formation dips steeply west and its strike is northwest at a slight angle to the fault contact that bounds it on the west.

Tightner Formation. The Tightner formation occurs in both the central portion of the Alleghany district and in the Downieville district. At Alleghany, the Tightner formation is in fault contact with the underlying Blue Canyon formation to the east and is conformably overlain by the Kanaka formation to the west.

The formation consists predominantly of massive beds of greenish hornblende-chlorite schist with varying amounts of quartz. Also present are minor amounts of carbonaceous slate, coarse white limestone, quartz, andalusite, and chlorite-mica schist.

Kanaka Formation. The Kanaka formation is exposed west of the Tightner formation and conformably overlies it. The Kanaka is exposed in a 1,600-foot-wide belt just west of Alleghany, on Oregon Creek west of Forest, and along State Highway 49 east of Goodyears Bar.

There are three members of the Kanaka formation. The basal member on the east is composed of a metamorphosed conglomerate of andesite and quartzite pebbles and cobbles in a matrix of black slate. Overlying the conglomerate is the chert member which consists of banded

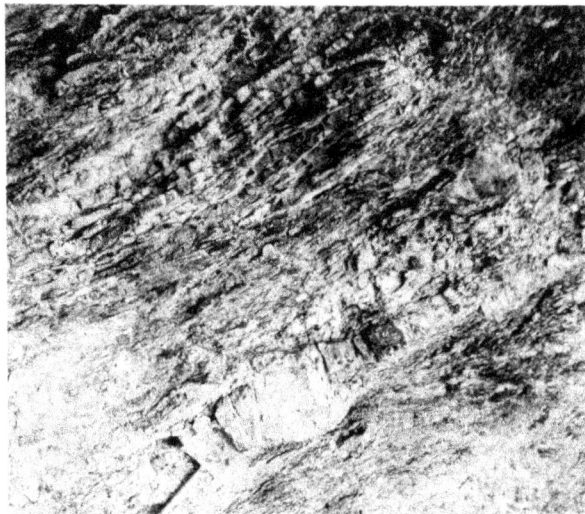

FIGURE 2. Road cut exposing Sixteen-to-One vein in conglomerate member of the Kanaka formation.

and contorted gray chert with interbedded black slate and fine-grained quartzite. The upper member is composed of successive beds of slate and greenstone.

Relief Formation. Small and scattered beds of siliceous metasediments lying between the Kanaka and Cape Horn formations in the Alleghany area are believed to be a north extension of the Relief quartzite (Ferguson and Gannett, 1932, p. 12). These beds are composed essentially of dark fine-grained quartzite and quartzitic mica schist.

Cape Horn Formation. The uppermost member of the Calaveras group in the area is the Cape Horn slate which occurs in a north-trending belt 1 to 2 miles wide along the western portion of the area. The formation is composed chiefly of dense black clay-slate with interbedded thin bands of light-colored volcanic material.

Igneous Rocks

Serpentine. Two large masses of serpentine extend along the eastern and western portions of the area, and many smaller masses lie between them. Serpentine is an alteration product of deep-seated ultrabasic igneous rocks of Jurassic (?) age. Serpentinization was prior to the formation of the quartz veins. Serpentine, if near a gold-bearing quartz vein, has a marked effect on the character of the ore deposit. (See section on ore deposits.)

The serpentine ranges from light to dark green in color and contains light-colored shear planes. Magnetite and occasionally chromite are present. Serpentine, under the influence of vein-forming solutions, has been replaced by carbonate (chiefly ankerite) and mariposite. Further alteration in places has produced talc.

Gabbro, Diorite, Quartz Diorite. Dark-colored granitic rocks ranging in character from amphibolitic rocks derived from gabbro to schistose diorite and quartz diorite are found in much of the area.

Extensive outcrops occur east of Downieville, south of Goodyears Bar, and immediately west and south of Alleghany. Numerous smaller masses are found in the area, expecially in areas occupied by the Tightner and Kanaka formations.

These intrusive rocks are composed chiefly of hornblende and alteration products of feldspar with smaller amounts of quartz, pyrrhotite, biotite, and olivine. The general trend of these rocks follows the strike of the older formations, but where contacts are exposed the older schistosity is cut by the gabbro. A post-Carboniferous (probably Jurassic) age has been assigned these rocks by most writers.

Medium and Light-Colored Granitic Rocks and Aplite. A moderate-sized mass of granite crops out 5 miles southwest of Goodyears Bar. A zone of small, altered granite stocks and dikes occurs in the central portion of the area. There also are a few small dikes of aplite in the Alleghany area. The large granitic bodies of the Sierra Nevada batholith lie approximately 5 miles east and 10 to 15 miles west of the area.

Neither the granite masses nor the aplite dikes are schistose. The granites are medium grained and are composed essentially of quartz and feldspar. The principal accessory mineral is biotite but it is generally not abundant. The aplite dikes are fine grained and are of similar composition to the granite.

The granites and their satellites are generally believed to be of Upper Jurassic or Lower Cretaceous age.

Tertiary Rocks

The oldest Tertiary rocks in the area are river gravel deposits of Eocene age. These are characterized by an abundance of quartz pebbles and are rich in placer gold. Overlying the Eocene gravels are Miocene intervolcanic river gravels. These deposits contain an abundance of volcanic rocks, mainly rhyolite and andesite.

Large masses of volcanic breccia of the Mehrten formation (Mio-Pliocene) cap the interstream ridges in the area. The breccia is composed of angular and water-worn fragments of andesite loosely set in a matrix of finer andesitic debris.

Dense fine-grained olivine basalt (Pleistocene), which represents a succession of flows, crops out in the area, notably on the summit of Bald Mountain at Forest. Pleistocene and Recent gravels occur in and near the present stream channels.

The Tertiary rocks invariably are gently dipping and lie with distinct angular unconformity upon the older stratified rocks.

Veins

The veins of the Alleghany area are quartz-filled fissures and replacement bodies in a fracture system that was formed principally by north-to-northwest-trending reverse faults. A few of the veins have followed probable normal faults. Dip is moderately east or west with the east-dipping veins tending to be flatter and more persistent. The veins are well defined, vary greatly in width, and are generally irregular in strike and dip. Average width of the major veins is from 5 to 6 feet, but there is considerable variation in vein dimensions.

The veins were formed prior to deposition of the Eocene auriferous gravel and after the intrusion of the granitic rocks. The veins are closely related to the granitic rocks and were formed near the end of

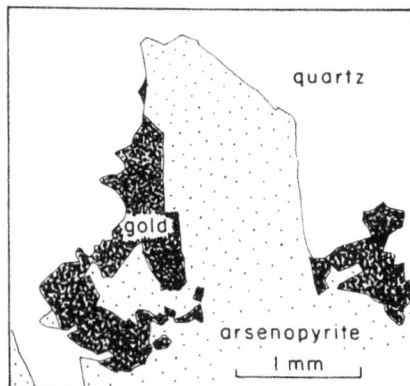

FIGURE 3.   Sketch showing gold re-
placing quartz and arsenopyrite (after
Ferguson, 1915, p. 165).

late Jurassic time. The veins were formed many thousands of feet below the land surface of that time.

Quartz veins occur in all of the pre-Tertiary rocks of the area except the larger serpentine masses. Generally, none of the veins penetrate the larger serpentine masses, but some abut against them. Where the fissures extend into the serpentine, the vein frays into small stringers which grade into a replacement mixture of mariposite and carbonate (chiefly ankerite).

Ribbon structure, which is so characteristic of the Alleghany veins, implies a banded or streaked appearance occasioned by slabs or thin sheets of country rock or other material interlayered with quartz in the veins. Some of the ribbons are straight and generally parallel to the walls while others are irregular. The irregular structures are known as crinkly banding. Carbonate material (chiefly ankerite) predominates in the vein filling while other common minerals are micas and sulfides. Ferguson and Gannett (**1932. p. 36**) **attrib**uted the formation of ribbon

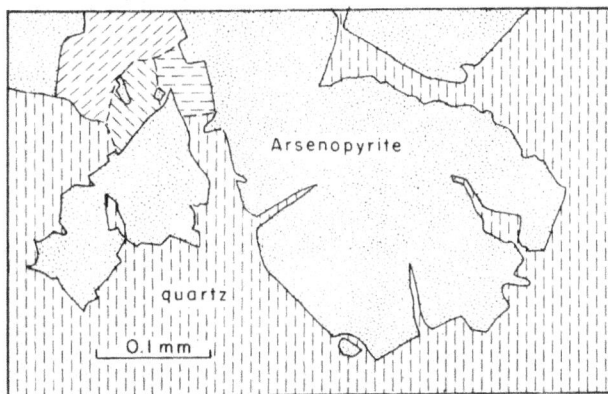

FIGURE 4.   Sketch showing arsenopyrite bordered and em-
bayed by a single quartz crystal, seen by polarized light.
The different quartz crystals are indicated by the directions of the broken lines. (After Ferguson, 1915, p. 167.)

quartz to accretion where later fracturing was followed by deposition of still later minerals along closely spaced planes. McKinstry and Ohle (1949, p. 108), however, conclude that ribbon structure results almost entirely from replacement. Cooke (1947, pp. 229-232) in a study of the Sixteen-to-One vein, believes that a combination of accretion and replacement was responsible for the ribbon structure.

The quartz veins of the Downieville area are fissure and replacement veins, ranging in strike from northwest to northeast and dipping moderately to steeply east. Width of the veins ranges from 2 to 10 feet. These veins have a higher sulfide content than the Alleghany veins.

Ore Deposits

The most characteristic feature of the ore deposits in the Alleghany district is the extreme richness, but very erratic distribution and commonly small extent of the ore shoots. Most of the gold produced from mines in the vicinity of Downieville has been from sulfide-rich ores whereas those at Alleghany contain a much larger proportion of free gold.

At Alleghany, the high-grade ore shoots range in size from small bunches yielding a few hundred dollars to those which have yielded hundreds of thousands of dollars. An ore body at the Sixteen-to-One mine, which was about 40 feet long, contained nearly $1,000,000 in gold while an area 14 by 22 feet as seen in section in the vein of the Oriental mine yielded $736,000 (Ferguson and Gannett, 1932, p. 52). Quart-vein matter outside of the shoots is nearly barren of gold and mill-head assays commonly yield $5 per ton or less in gold. Although the quartz adjacent to a high-grade shoot generally is barren, it has been found that the change may be less abrupt in one or more directions than the others, and along these courses the quartz may contain more gold than the average vein quartz outside of the shoots (Ferguson and Gannett, 1932, p. 52).

At Alleghany the high-grade ore rarely extends across the width of the vein, but is usually confined to strands near the footwall or the hanging wall. Veins with an easterly dip have been the chief source of gold in the area. However, in the past few years west-dipping veins at the Brush Creek and Kate Hardy mines have proved important.

There are certain favorable structural features associated with high-grade ore deposits in the Alleghany district. Any irregularity in the vein caused by changes in dip and strike, sudden swellings, junctions or splits in the vein, and minor faults are the principal favorable locations for the deposition of high-grade ore shoots. The influence of large serpentine masses is of importance as the veins tend to bend toward parallelism with the serpentine contact. Much of the production of mines near serpentine come from that part of the vein within the bend. Smaller serpentine dikes appear to have some influence also on localization of high grade ore.

The presence of visible free gold is of great importance in tracing shoots of high-grade ore, as there is commonly at least one direction in which the change to barren quartz is not as sharp as in the other directions. Coarsely crystalline arsenopyrite is also of importance in finding ore shoots as this mineral seems to have been effective in causing the deposition of gold. In some places the arsenopyrite is partly replaced by gold (Ferguson and Gannett, 1932, p. 59). The presence of mariposite or "blue-jay" is sometimes significant as it is most abundant near serpentine, and the presence of that rock has had considerable control over the location of high-grade ore shoots.

In the vicinity of Downieville, a higher proportion of the gold occurs in relatively large, but lower grade ore shoots as compared with those in the Alleghany area. The gold occurs in quartz veins, and occasionally in contact metamorphic rocks closely associated with sulfides, chiefly pyrite. Subordinate amounts of arsenopyrite, galena, chalcopyrite, and sphalerite are commonly present. Ore produced from the Gold Bluff mine in 1938 averaged $10 per ton in gold (Averill, 1942, p. 28), while ore from the Oxford mine ranged in value from $10 to $15 per ton (Huelsdonk, L. L., personal communication, 1954).

Mineralization

On the basis of studies made on thin sections, Ferguson and Gannett (1932, pp. 38-51) have divided mineralization in the Alleghany area into four stages—the chlorite, quartz, carbonate, and final stages.

The earliest stage of mineralization is the chlorite stage. This stage, which was probably on a regional scale, consisted essentially of the chloritization of the hornblende of the schists and altered gabbro. Partial serpentinization of the peridotite and pyroxenite belong to this stage.

The quartz stage followed, during which fissure filling by quartz and quartz replacement of the wall rock occurred. Radiating crystals of arsenopyrite, which formed the nucleus for later deposition of native gold, and pyrite were deposited in the fissures and veins, both slightly earlier than and contemporaneous with the quartz. Other minerals formed during this stage were plagioclase, barite, apatite, and beidellite in small amounts.

The carbonate stage followed the quartz stage. During this, there was extensive replacement of the wall rocks by carbonate (chiefly ankerite), sericite, mariposite, and minor amounts of chlorite, arsenopyrite, and pyrite. Some earlier quartz was partially replaced. In the fissures, native gold, along with sericite, carbonate, mariposite, and several sulfide minerals, replaced the earlier quartz. The earlier brecciated quartz of the quartz stage was recemented by later quartz of the carbonate stage. Economically, the carbonate stage was the most important of all the stages.

In the final stage, small amounts of uneconomic minerals including calcite, finely crystalline pyrite, and possibly marcasite, were deposited as small veinlets or as coatings on drusy quartz.

MINING AND MILLING

As the high-grade ore of the Alleghany gold mines is sporadic in distribution a close correlation between favorable geologic conditions and development work must be exercised. Ore bodies are easily missed unless the veins are thoroughly explored. The discovery of ore shoots and isolated bunches of ore, as practiced now, is made by cutting up the more favorably situated portions of the vein into small blocks by means of drifts and raises, and stoping the ore thus blocked out. The mining of large amounts of nearly barren quartz to avoid missing high-grade bunches is thus avoided.

Most of the lode-gold mines of the Alleghany-Downieville area are worked through adits and winzes on the veins. A few mines were originally developed by inclined shafts from the surface although none are used at the present time. Levels are generally established at intervals of 100 to 200 feet. The ground between the levels is developed by raises and often intermediate drifts.

Underground excavation entails drilling, blasting, and timbering as required. Ingersoll-Rand stopers and Swedish-Atlas Copco jackleg drills are most commonly used in mining and development work. In some cases stopers mounted on air bars or rail-mounted hydraulic jibs are utilized in drift and crosscut headings. Bits equipped with tungsten-carbide inserts are commonly used. Slushers are used for handling the muck and ore in intermediate and flat raises and pneumatic mucking machines on the drift and crosscut headings.

Very little support timbering is required in the underground workings except where the country rock consists of serpentine. Because of the tendency of serpentine to swell upon exposure to air, much timbering is required.

Because most of the mines are developed by adits and underground shafts, underground hoists are used. Tramming of the ore cars generally is done by storage-battery locomotives.

Milling-grade ore in the area is processed by crushing and grinding, gravity concentration, amalgamation, and flotation. Mill head assays range from less than 5 dollars to 30 dollars or more in gold per ton. Mill recoveries range from about 80 to 96 percent.

Gold ore is first crushed by a jaw crusher and then wet-ground in a ball mill. Stamp mills were once used in secondary crushing, but their size, power consumption, and other economic factors, as compared with those of ball mills, have rendered them obsolete. None are used in the mills of this area. Jaw crushers reduce the ore to a convenient size for the ball mill feed. Particle sizes coming from the jaw crushers range from minus-$\frac{3}{4}$ to minus-2 inches while ball mills further grind the ore to sizes ranging from minus-6 to minus-65 mesh for tabling, and to minus-100 mesh for flotation.

The finely ground ore is then sent to gravity-concentrating devices, including jigs and hydraulic traps. Overflow from the jigs and hydraulic traps flows into spiral or rake classifiers. Sands from the classifier are usually reground in the ball mills, and introduced again through the mill circuit. Slime products from the classifier may be sent to either concentrating tables or to a flotation unit. Scalping screens are utilized in the mill circuits in certain cases.

Concentrating tables, which include several types, are shaking tables, each equipped with a series of riffles. Water is allowed to flow down the slope of the tables and aids in gravity separation. Several tables are usually used in series so as to minimize the gold loss from the fine particles. Concentrates from the tables generally are sold to the smelter at Selby.

To minimize a further gold loss, particularly where the gold is in intimate aggregation with the sulfide minerals, the slimes are directed into flotation cells. The gold-bearing sulfide concentrates produced from flotation are usually sold to the smelter at Selby or the Empire Star mine in Grass Valley. In one instance, the flotation concentrates are sent to a thickener for dewatering.

High-grade gold ore is handled separately and not processed in the mill. Such ore is crushed in laboratory-sized rod mills, gyratory crushers, or rolls. The gold is amalgamated in amalgamating barrels or Berdan pans, retorted, and the bullion is sold to the U. S. Mint in San Francisco.

MINES ACTIVE SINCE 1942

With the onset of World War II, high wages, scarcity of materials, and the departure of men to engage in defense activities or to serve in the armed forces, caused gold mining in this region to be severely curtailed. War Production Board Order L-208, which was issued on October 8, 1942, in effect ordered all gold miners to go into other types of mining. Most of the gold mines of the Alleghany-Downieville area were closed during the war. The Sixteen-to-One mine was operated on a limited basis under a special grant issued by the War Production Board.

Order L-208 was lifted on July 1, 1945. The Sixteen-to-One mine resumed normal operations soon afterward, and in 1946, the Brush Creek and Oriental mines were reopened. Lode gold mining in this area, however, has not been resumed on a scale equal to that before World War II.

Lode mines which have had the greatest amount of activity and production since the war are the Sixteen-to-One, Oriental, Brush Creek, Kate Hardy, and Red Star mines. Others in which there has been some activity and production include the El Dorado, Eureka, Gold Bluff, Gold Crown, Irelan, Mugwump, Oxford, and Seven Aces mines.

Brush Creek Mine. Location: Secs. 17, 18, 19, T. 19 N., R. 10 E., M. D. M., 1½ miles south of Goodyears Bar. Ownership: Alpha Hard-

FIGURE 5.   Ore train entering Cassidy level, Brush Creek mine; camera facing south.

ware & Supply Company, F. F. Cassidy, Nevada City. Leased by the Best Mines, Inc., P. O. Box 177, Downieville, California. Company officers are: Mrs. Irene Best, president, B. C. Austin, vice president, and L. L. Huelsdonk, secretary-treasurer and general manager. W. Reed is mine foreman and J. Folsom is mill foreman.

By 1875, an estimated $1,000,000 had been produced from workings of the old Brush Creek shaft (Ferguson and Gannett, 1932, p. 87). In 1922-23, the Ante-Up Mining Company drove the Brush Creek tunnel, which intersected the old shaft, sunk a winze, and produced a few thousand dollars worth of gold (Logan, 1929, p. 160). In 1927, the Kate Hardy Mining Company did some additional development work in the old workings.

The mine lay idle until reopening in 1944 by A. L. Merritt, who operated it until March 1950. The mine was then leased by the Best Mines Company. Until 1951, the mine was worked through the Merritt adit and the old Cassidy shaft. In 1951 the Cassidy adit, which is now the main haulage way for the mine, was started. The mine is active at the present time, and to date has a total production of nearly $4,000,000, half of which is based on a price of $20.67 per ounce (L. L. Huelsdonk, personal communication, 1954).

The quartz vein in the present workings ranges from 3 to 15 feet thick, strikes north, and dips west at an average of 42 degrees. The vein occurs near and along the contact of black slate of the Cape Horn formation to the west and serpentine to the east. In some places it lies largely within the slate, but in others serpentine forms the footwall and slate forms the hanging wall. Native gold occurs sporadically as high-grade concentrations in quartz and is associated with minor amounts of auriferous arsenopyrite, galena, and pyrite. One of the most recent high-grade ore discoveries was on the 125-foot level off the Golden Gate shaft. The ore shoots commonly are nearly vertical. A replacement mixture of carbonate and mariposite attains a thickness of up to 50 feet between the vein and the serpentine footwall. The gold averages 830 in fineness.

The main working entry of the Brush Creek mine at present is the 2,600-foot Cassidy adit which extends in a south to southwest direction. The first 1,300 feet of the adit is in serpentine while the remainder is in Cape Horn slate. Elevation of the adit portal is 3,158 feet. The Merritt Crosscut adit, 1,500 feet south of the Cassidy adit portal, and 171 feet higher in elevation, extends 600 feet west to the vein and is the other working entry. Three raises connect the Cassidy and Merritt levels from which intermediate drifts and stopes have been worked. Nearly all of the ore production since World War II has been from this portion of the mine.

During the summer and fall of 1954, a winze known as the Golden Gate shaft was sunk from the Cassidy adit 2,000 feet in from the portal. On the 125-foot level of this winze, a 60-foot southeast crosscut was driven to the vein and 30 feet of drifts driven on the vein to October 1954. Present production is from this level and from the Cassidy adit which is being extended.

Older workings of the Brush Creek mine which are not used at present include the 60-foot inclined Brush Creek shaft, the 2,200-foot Brush Creek adit which intersects the shaft, the 5,000-foot Extension adit, the 600-foot Ante-Up adit, and the 500-foot Peavine adit. The collar

FIGURE 6.   Mine yard at Cassidy adit, Brush Creek mine. Loading bins are shown at right. Camera facing north.

of the Brush Creek shaft is just south of the Mountain House saddle, 1½ miles south of the Cassidy adit portal.

Five stopers, which include three Ingersoll-Rand 48 and two Ingersoll-Rand 58 and two T-350 Sullivan drifters mounted on Joy jumbo arms are used for drilling. Ingersoll-Rand screw-on bits with tungsten inserts are used. The mine is equipped with two 500-cubic foot Ingersoll-Rand compressors. The mine workings are ventilated by a 350-cubic foot blower located on the Cassidy level. Two storage battery locomotives, a Mancha and a General Electric, are used for tramming the ore trains. Capacity of the ore cars used on the Cassidy level are two tons while those used on the Merritt level are one ton. An underground hoist in the Cassidy adit serves the Golden Gate shaft.

High-grade ore is hand sorted at the place where it occurs and is handled and milled separately. Milling ore and waste are trammed from the two adits and deposited into respective bins. The ore is trucked 5 miles to the mill, which is located at the Oxford mine, 1 mile north of Downieville.

About 30 tons of ore are treated at the mill each day. The millheads average $30.00 per ton in gold (L. L. Huelsdonk, personal communication, 1954). The mill consists of a primary jaw crusher, secondary ball mill, Miners Foundry jig, Huelsdonk table, Denver grinding pan, a Wemco Dorr-type thickener, Wemco rake classifier, and Denver sub-A flotation cells. The free gold is amalgamated, retorted and sold to the mint. The flotation concentrates are sold to the Empire Star mine in Grass Valley.

There are 22 men working at the mine and two men at the mill. The work schedule is a 6-day week, and generally one shift per day.

Gold Bluff Mine. Location: secs. 23 and 26, T. 20N., R. 10E. and secs. 2 and 3. T. 19N., R. 10E., M. D. M., 1½ miles north of Downieville on the Downie River. Ownership: Mrs. Irene Best, Downieville.

Originally worked in 1851, the Gold Bluff gold mine was active almost continuously until 1902 (MacBoyle, 1920, p. 10). The mine was again in operation during 1914-1916. In 1948, the Best Mines Inc. purchased the property and, in 1950, the mine was rehabilitated and several hundred feet of drifts and crosscuts were driven. Several hundred tons of ore were produced, largely from support pillars remaining in the old workings. The ore was treated at the Oxford mill. Except for occasional maintenance work, the mine is idle at the present time. Total production for the Gold Bluff mine is estimated to be $1,500,000 based on a gold price of $20.67 per ounce (L. L. Huelsdonk, personal communication, 1954).

Native gold and auriferous pyrite occur in two quartz veins, the Fault or Gold Bluff and Crescent veins. The Fault vein strikes N. 20° W., dips 70° northeast, and ranges from 3 to 5 feet wide. The Crescent vein averages 5 feet wide, strikes N. 60° E. and dips 45° northwest. The veins occur at or near the contact of metavolcanic rocks and slate. The metavolcanic rocks lie to the east.

The mine is developed by two adits, the No. 2 and C. L. or No. 4 adits which crosscut west to the veins. The C. L. or No. 4 adit, the portal of which is located on the Downie River, was the main haulage way. It extends 1400 feet west to the veins and from there, drifts extend north. A 500-foot north inclined winze was sunk from the No. 4 level about 200 feet north of the No. 4 adit. The No. 3 adit, which is 210 feet above the No. 4 adit, extends 1,050 feet west to the veins. A 600-foot south-west crosscut was driven on the No. 4 level southwest of the veins by the Best Mines Company in 1950.

At the present time there is no equipment on the mine property except for several storage sheds and cabins near the C. L. adit portal.

Gold Crown (Wonder) Mine. Location: sec. 3, T. 18N., R. 10E., and sec. 34, T. 19N., R. 10E., M. D. M., in Alleghany. Ownership: Gold Crown Mining Corporation, Alleghany.

The Gold Crown gold mine was first worked as the Wonder mine. By 1922, the Flynn Brothers, H. T. Bradbury, and the J. Walsh Estate had done considerable development work (Logan, 1922, p. 519) in the old upper tunnel workings. In the middle 1920's, the Sierra Wonder Company drove a lower tunnel to the serpentine contact. In the latter 1930's the Duke family operated the property as the Wonder gold mine and started a new upper tunnel above the Flynn and Bradbury workings and began sinking a shaft.

In 1944-45, the Wonder Mining Company deepened the Duke shaft to 67 feet. The property was idle until 1949-50 when the company was re-incorporated as the Gold Crown Mining Corporation. During this time the shaft was deepened and some work was done on the north 100-foot level and the 67-foot level south of the shaft.

Between July 1951 and July 1954, the Giles brothers, development contractors for Gold Crown Mining Corporation, continued development work on the 100-foot level and in the lower tunnel. Since 1952, all development work was confined to the lower tunnel. The mine is intermittently active at the present time.

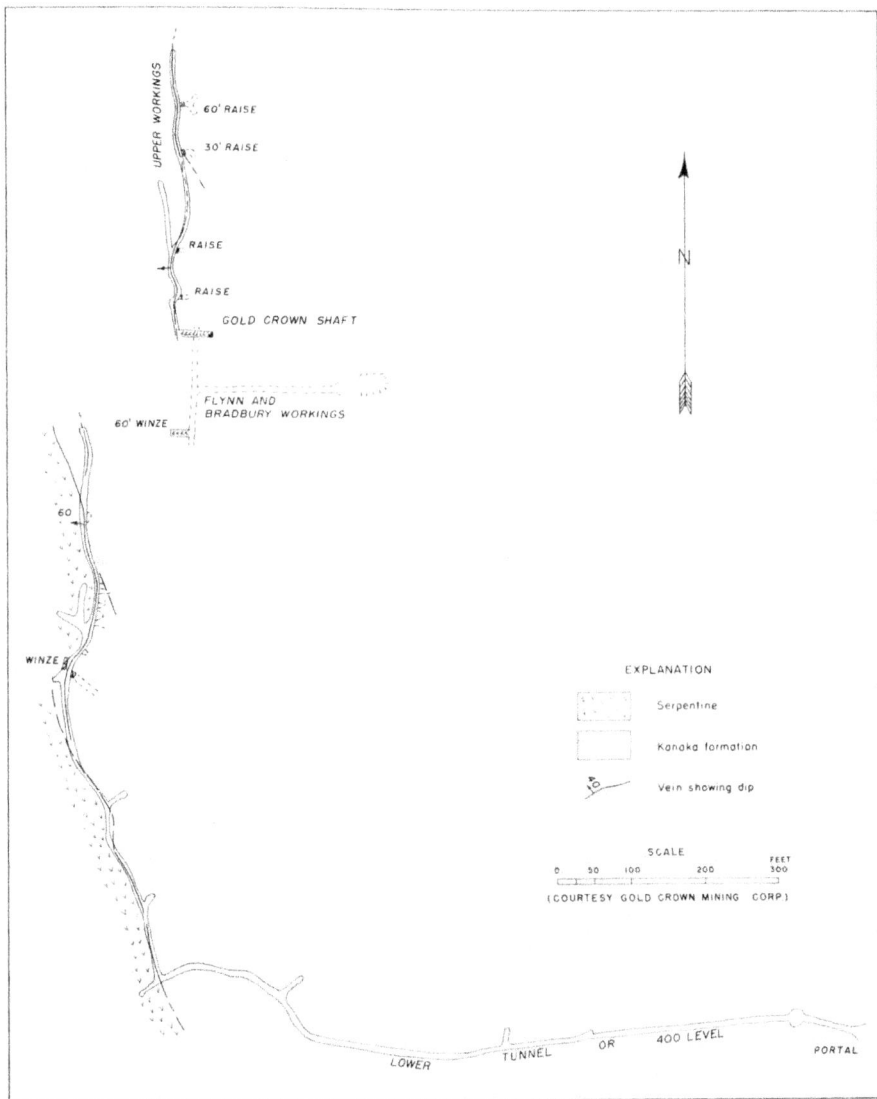

FIGURE 7.   Underground workings of Gold Crown mine.

In the lower level, the vein was encountered by exploring the serpentine-Kanaka formation contact. This vein diverged from the serpentine-Kanaka formation contact at a small angle and pinched out against the serpentine. The vein was then followed about 300 feet to the north. The vein has a general strike of north, dips west, and ranges from a few inches to 4 feet in width. Some high-grade ore was found near the serpentine. County rock includes serpentine and interbedded slates and tuffs of the Kanaka formation.

Mine development work includes a 104-foot shaft, a south drift, crosscut and winze on the 67 level and 470 feet of drifting and two small raises on the north 100 level. Other development work consists of a 200-foot west crosscut to the vein and drifts north 100 feet and south 75 feet. A 60-foot winze was sunk on the vein in the south drift 60 feet from the crosscut (Logan, 1922, p. 519). These upper workings are not being used at the present time.

Present development work is in the main adit or lower tunnel, the portal of which is about 0.3 mile southeast of the shaft and about 520 feet lower in elevation than the collar of the shaft. The lower tunnel was crosscut 1,000 feet west to the serpentine-Kanaka formation contact. Then the development work continued along the contact 400 feet north where a west dipping, north-striking quartz vein was encountered. The vein has been drifted on for a distance of 300 feet and some stoping has been done. A winze is being sunk from the lower level on the vein near the serpentine. During the middle of 1954, the winze had been sunk to a depth of 6 feet.

Swedish-Atlas Copco drills are used for drilling rounds. A mucking machine is used to load ore into 1-ton side-dump ore cars. Milling-grade ore is stockpiled at the present time as there is no mill on the property.

In the middle or Hyland level, the main vein strikes N. 45° W., while in the lower level, the strike averages N. 40° W. Two branch veins diverge to the north in the lower level 500 feet in from the portal. Dip of the vein ranges from 30 to 70 degrees northeast; the dip of the vein in the vicinity of the No. 1 ore zone, however, averages less than 40 degrees.

The No. 2 ore zone is located just northwest of the adit in the upper level and near the end of the lower level. Recent production was obtained from the No. 1 ore zone in the Hyland level between 300 and 500 feet in from the crosscut adit. None of the levels are connected underground.

Country rock consists of highly sheared gabbro, tuff, and slate of the Kanaka formation. Serpentine is encountered in the hanging wall in the Hyland level about 600 feet in from the portal. The productive parts of the vein are associated with carbonate alteration of the wall rock. Native gold occurs sporadically in bunches and is nearly always associated with coarse arsenopyrite.

Light-weight Swedish-Atlas-Copco jackleg drills are used in mining. The ore is hand trammed to a storage bin near the portal from where it may be trucked to a mill. Surface equipment at the mine includes a 10-by-12-inch Gardner Denver compressor, boarding house, blacksmith shop, and storage sheds. Five men are employed at the mine.

FIGURE 8. Kate Hardy mine and mill, from south side of Oregon Creek;
camera facing west.

Kate Hardy Mine. Location: sec. 29 and 32, T. 19 N., R. 10 E.,
M. D. M., 2½ airline miles northeast of Alleghany. Ownership: H. E.
Hawn, P. O. Box 748, Grass Valley, California.

The Kate Hardy mine was worked at intervals between 1860 and
1927. Between 1941 and 1953, the mine was leased and worked by the
O'Donnell brothers. In April, 1953, H. E. Hawn purchased the mine
from the Beggs Estate and some development work, mining and milling
have been done since that time. Total gold production for the mine is
approximately $547,000 (H. E. Hawn, personal communication, 1954).

The quartz vein, which occupies a reverse fault, has an average
strike of N. 23° W. and dips 75°-80° southwest. The quartz vein
ranges from a few inches to several feet in width, although a maximum
width of 55 feet has been encountered. The vein lies in the Cape Horn
slates and cuts the slate at a small angle in strike and dip (Ferguson
and Gannett, 1932, p. 90). Dikes of gabbro and serpentine cut the vein
irregularly in places. Considerable slate has been replaced by carbonate
close to and within the vein. Mariposite is erratic in distribution and
is not necessarily confined to the vicinity of the serpentine. Arsenopyrite, pyrite and galena are present.

The principal workings are south of Oregon Creek and consist of
several adits, the lowest a few feet above the creek level (No. 4 tunnel)
and the highest 250 feet above the creek. A 75° inclined winze was
sunk from the No. 4 level about 300 feet in from the portal. Three
levels branch off the winze north and south, the principal level being
the No. 1 level which is about 100 feet below the No. 4 level. The No. 1
level extends to the north about 1,000 feet. Most of the production
previous to 1927 came from the south workings.

Since 1940 most of the gold production has been from the workings north of Oregon Creek. The north workings include lower and upper drift adits. The lower adit is 66 feet lower in elevation than the upper adit and was driven a distance of 1,350 feet along the vein. A raise, which was driven about 500 feet in from the lower adit portal, extends to a height of about 44 feet. From the top of the raise, an intermediate drift was driven 35 feet south. At this point a raise was driven about 22 feet to connect into the upper adit. The upper adit was started in 1951.

During 1954, the upper adit was extended to its present 600-foot length, the two north adits were connected underground, and a raise off the upper adit was driven to a height of about 70 feet. At the present time, mining and reopening work is underway on the No. 1 level below the north workings. An Ingersoll-Rand 48 stoper is used for drilling and the ore is hand-trammed to the mill, which is located immediately west of the north adit portal.

Small tonnages of ore are milled at the present time. The mill, which has a rated capacity of 40 tons per day, is equipped with a 3 x 4-foot ball mill (which produces a minus 20-mesh product), Miners Foundry jig, Dorr rake classifier, Knudsen Bowl, and Wilfley tables. The free gold is amalgamated, retorted, and sold to the U. S. Mint in San Francisco. Concentrates are shipped to the Empire-Star mill in Grass Valley. Three men are employed at the mine.

Irelan (Ireland) Mine. Location: secs. 10 and 11, T. 18 N., R. 10 E., M. D. M., 2 airline miles south of Alleghany on the south side of Lafayette Ridge. Ownership: A partnership known as the Yuba-Irelan Mines, Ltd., Joseph Harris, manager, Alleghany.

The Irelan mine is the only active gold mine in the Lafayette Ridge area at the present time. The mine was originally worked around 1860 and was active at intervals until 1924 (Ferguson and Gannett, 1932, p. 127). During the 1930's the principal workings were extended, but there was little ore production. In 1936, a fire destroyed most of the surface plant. In 1949, George Hyland purchased the mine and developed the Hyland level. Since 1953, the mine has been operated by Yuba-Irelan Mines, Ltd. The lower level is being rehabilitated while some ore has been produced from stopes in the Hyland level. This ore was milled at the Red Star mill in Alleghany during July and August, 1954. The mine is under development at the present time. Total estimated production for the Irelan mine is slightly more than $500,000 based on a gold price of $35 per ounce (Joseph Harris, personal communication, 1954).

There are two quartz veins, the east and west on the property; they trend northwest and dip moderately to steeply northeast. Virtually all of the mine production has been from the east or main vein. In the upper level or Waldron level, the main vein has an average strike of N. 60° W., but changes sharply to N. 30° W. about 500 feet in from the crosscut adit and continues on this course for 500 feet. At this point the main vein bends toward its original direction again. The change in direction corresponds to a projected junction with the west vein (Ferguson and Gannett, 1932, p. 128) and the greatest amount of production has been from this segment of the vein which is known as the No. 1 ore zone. Much of the vein in the vicinity of the No. 1 ore zone has been stoped out.

Mugwump (Young America) Mine. Location: secs. 28 and 33, T. 19 N., R. 10 E., M. D. M., a quarter of a mile southwest of Forest. Ownership: Cecil Vivian, Forest.

Originally known as the Young America mine, the Mugwump mine was first worked in 1852 (Ferguson and Gannett, 1932, p. 93). It is a combination lode and drift mine, but the greatest amount of development work has been in the buried gravel channel with the majority of the total mine production being placer gold. The quartz vein was encountered in bedrock when the buried channel was originally developed. The property was active from 1892 to 1896 and again from 1916 to 1928 (Logan, 1929, p. 195). Since November, 1953, the present owner has been developing the mine through a new adit. Estimated lode production is about $5,000 at a gold price of $20.67 per ounce (Ferguson and Gannett, 1932, p. 93).

The vein occupies a fault zone in slate and green schist of the Kanaka formation. The vein strikes N. 35° W., dips 60 to 72 degrees southwest, and measures as much as 3 feet in width. Small but high-grade ore shoots composed of native gold associated with arsenopyrite were found in the winze just below the adit level and from an area 850 feet to the southeast near the end of the lower level. In both places where high-grade ore was found, the dip of the vein was flatter than usual (Ferguson and Gannett, 1932, p. 94). A north-striking vein some 1,300 feet to the northwest of the main vein and exposed about 100 feet in from the adit portal may be a faulted segment of the main vein.

The older workings are developed by a southeast-extending adit, the portal of which is on Oregon Creek. The vein is encountered 1,400 feet in from the portal. An inclined winze extends 125 feet to the lower level from this point. The main adit level drifts along the vein and then continues on to the southwest along the buried gravel channel. The lower or winze level drifts along the vein for about 850 feet southeast from the winze.

The new adit, the portal of which is about 600 feet southwest of the Mugwump portal, extends 800 feet southeast. It is entirely in green schist of the Kanaka formation. A raise has been driven to the gravel channel 600 feet in from the portal and a washing plant is being installed. The adit is continuing to be developed to the southeast toward the vein. Six men including Mr. Vivian work at the mine.

Oriental Mine. Location: sec. 4, T. 18 N., R. 10 E., and sec. 33, T. 19 N., R. 10 E., M. D. M., 1 mile southwest of Alleghany. Ownership: Drescher Estate, Financial Building, Sacramento, and Mrs. W. E. Kliensorge, 2161 Hyde Street, San Francisco; leased by The Dickey Exploration Company, Alleghany, Donald R. Dickey, general manager.

More than $2,000,000 was produced from the Oriental gold mine prior to 1890 (Ferguson and Gannett, 1932, p. 99). The mine was active in 1900 (Lindgren, 1900, p. 8) and again from 1915 to 1917 (MacBoyle, 1920, p. 108). A few thousand tons of low-grade ore was produced during the period 1924 to 1926 (Logan, 1929, p. 168). From the latter 1930's to 1942 the mine was active. The Dickey Exploration Company reopened the mine in 1946 and at the present time, production is ap-

FIGURE 9. Oriental mine yard; Oriental adit in center of photo, camera facing north.

proximately 30 tons of ore per day (Donald R. Dickey, personal communication, 1954). Estimated total production for the mine is $4,250,000 figured at a gold price of $35 per ounce.

The Oriental vein lies in the plane of a reverse fault that has a displacement ranging from 120 to 200 feet (Ferguson and Gannett, 1932, p. 99). The strike of the vein is west in the eastern portion of the mine. As the vein approaches the serpentine to the west the strike swings to N. 65° W. Dip ranges from 28 to 50 degrees north and northeast, the flatter dips being near the serpentine. Width of the vein ranges from thin stringers to 20 feet. Vein branching is common throughout the workings and near the serpentine, the vein branches into a complex series of small veins and stringers which grade into a replacement mixture of carbonate and mariposite.

The vein has been most productive near the serpentine-gabbro contact in the upper workings and in an area between the 400-foot level and the surface. In this latter area, the vein splits into several branches, and stopes following these branches from the 400-foot level to the surface yielded over $1,500,000 (Ferguson and Gannett, 1932, p. 100). The high-grade ore is nearly always associated with coarse crystalline arsenopyrite.

At the 8th level, a junction is made between the Oriental and Alta veins. The Alta vein is a flatter vein in the footwall of the Oriental vein. The strike is roughly parrallel to that of the Oriental vein and the dip is gentle to the north (Averill, 1942, p. 33).

A new vein was discovered in 1952, in the hanging wall, that is generally parallel to the main Oriental vein. Recent production has been concentrated close to the junction of this vein with the serpentine.

Schistose gabbro is the principal wall rock of the veins in the Oriental mine. In the western portion of the workings, serpentine forms the footwall. A mass of granite forms the wall rock of the vein extending from the 9th level to below the 19th level. On the 13th level, the granite extends for a distance of 350 feet along the vein. The Oriental mine is the only mine in the Alleghany area where the wall rock is granite. Other rocks in the general area include diorite, quartz diorite, and quartzite.

The mine is developed by a 3,800-foot crosscut known as the Oriental adit or 13th level at an elevation of 3,620 feet. This adit, which was driven N. 30° E., is the main haulage way for the mine. Near the end of the adit, drifts extend several hundred feet east and 750 feet west-northwest along the Oriental vein. An old inclined shaft, the collar

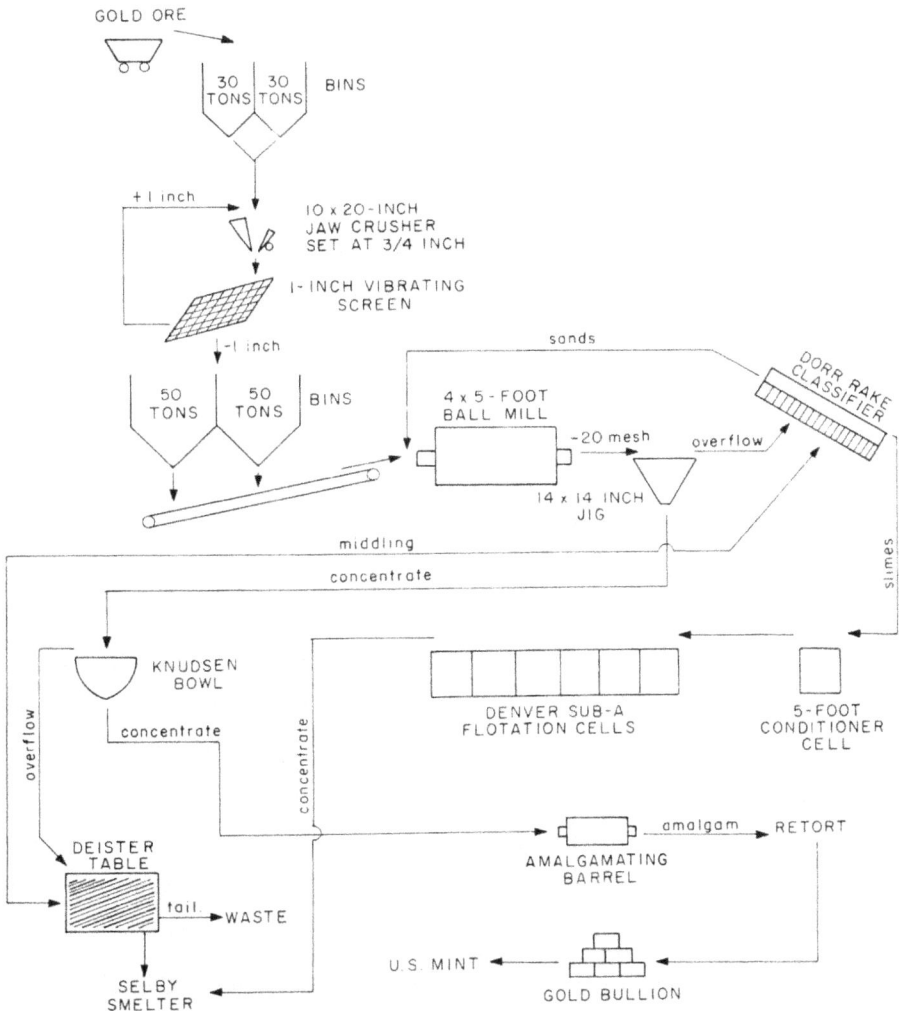

FIGURE 10.   Oriental mill flowsheet.

of which is 780 feet above the adit portal, was sunk on the vein. Ten levels ranging from 30 to 100 feet apart in elevation extend east and west along the vein from the shaft, the most extensive being the 4th level which is about 600 feet long. The shaft bottoms on the 12th level and a raise driven from the 13th level to the 12th level connects the shaft with the lower workings.

Since World War II, a 600-foot inclined winze has been sunk from the northwest end of the 13th level to a new level known as the 19th level. Drifts and crosscuts extend both northwest and southwest from this level. Most of the post World War II production has been from stopes between the 19th and 13th levels northwest of the winze and from the newly discovered hanging wall vein.

Ore is hoisted up the winze, and trammed by a Mancha electric locomotive along the 13th level to the mill. Lightweight Swedish Atlas Copco drills and mucking machines are used in mining.

Milling-grade ore is treated by crushing, grinding, jigging, and flotation. High-grade is treated separately by grinding and amalgamation.

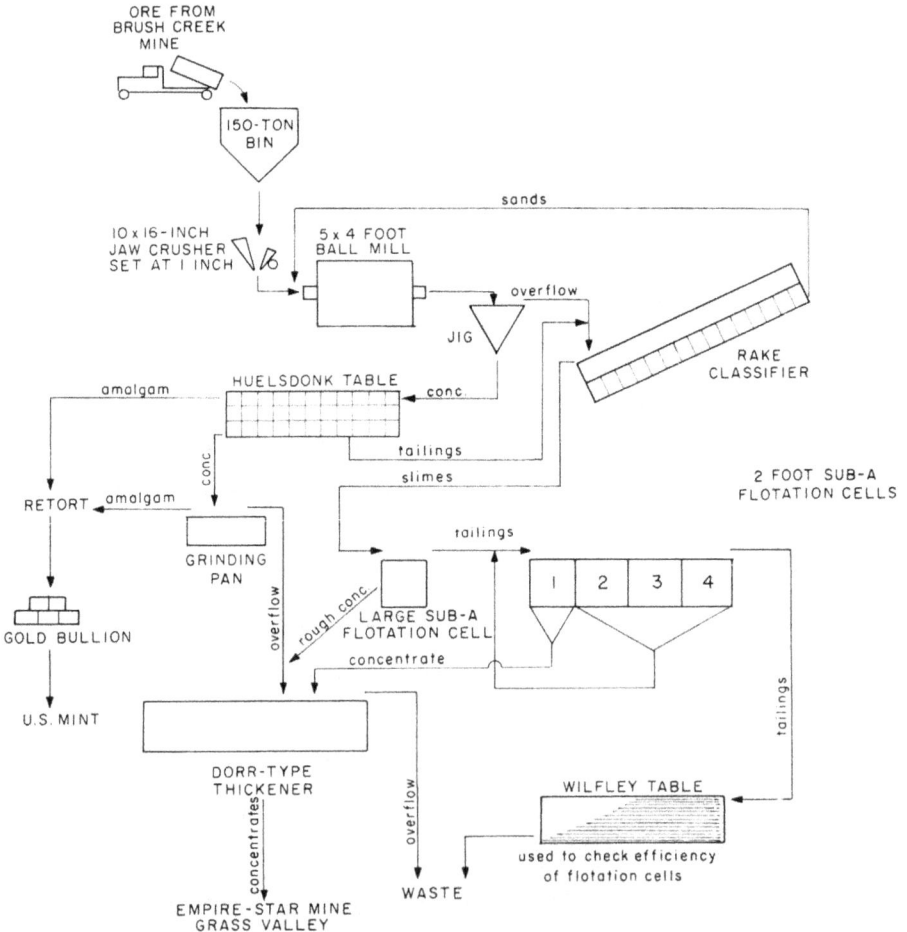

FIGURE 11. Oxford mill flowsheet.

Mill recovery ranges from 85 to 90 percent (Donald R. Dickey, personal communication, 1954). Mill capacity is rated 75 tons per day although 30 tons are milled per day at the present time. Eight men work in the mine and three men in the mill.

Oxford Mine. Location: secs. 22, 23, 26, 27, T. 20 N., R. 10 E., M. D. M., one mile north of Downieville. Ownership: Oxford Quartz Mining Company, Downieville.

The Oxford gold mine has been intermittently active since the early 1890's. Between 1945 and 1948 the Best Mines Company leased the property, during which time development work including some diamond drilling was done. The mine has been idle since 1948. (L. L. Huelsdonk, personal communication, 1954).

The principal vein strikes north, dips 72° east, and ranges in width from 2 to 7 feet. Country rocks consist of slate of the Calaveras group (Carboniferous in part). A body of serpentine was encountered in the Oxford adit about 50 feet west of the vein.

The Oxford adit (El. 3437 feet) is a west crosscut which extends through slate for a distance of 657 feet to the vein. At the vein, a drift extends for about 2000 feet to the north. The Snyder No. 2 adit is 493 feet above and about 1700 feet north of the Oxford level. The Snyder No. 2 level is a west crosscut extending through slate to the serpentine contact for 1100 feet. A 600-foot drift branches to the north from this crosscut.

The Snyder No. 1 portal is 183 feet above and 500 feet northwest of the Snyder No. 2 portal. The Snyder No. 1 adit crosscuts to the west for 50 feet. The end of this level is connected to the Snyder No. 2 level by a raise.

Although the mine is idle the mill on the property is leased by the Best Mines Company to treat ore from the Brush Creek mine.

Red Star (Osceola, Yellow Jacket). Location: sec. 34, T. 19N., R. 10E., M. D. M., 1 mile northwest of Alleghany. Ownership: Tightner Mines Company, 58 Sutter Street, San Francisco, California.

This property has been operated intermittently since the early days of the district. The property is a consolidation of two mines, the Osceola on the east and the Red Star on the west. The Red Star mine is part of the original property of the Tightner Mines Company, which was consolidated with the Tightner workings in 1911 (Robert McCulloch, personal communication, 1955). It was active from this date until 1925. During the 1930's the Osceola portion of the property was worked. The first 1,500 feet of the working entry to the Red Star mine is on the Osceola claim.

From 1942 until early 1954, the mine was leased and operated by the Yellow Jacket Consolidated Mines, Ltd. Late in 1954 it was subleased for a short time by a partnership consisting of C. J. Ayres, D. B. Morton, J. Bach, and H. J. Adams. The mine is idle at the present time.

Prior to 1942 two veins were developed on the property; the Red Star and Osceola. Both veins trend northwest and dip moderately to steeply northeast. The Red Star vein lies approximately 1,200 feet west of the Osceola vein. Since 1942, a new vein has been developed

22 | 23

27 | Sec. 26   T 20N R 10 E

SNYDER NO. 1
EL. 4097'

SNYDER NO. 2
EL. 3830'

Serpentine

N

SCALE

0   100   200   400   600 FEET

Serpentine

OXFORD PORTAL
EL. 3437'

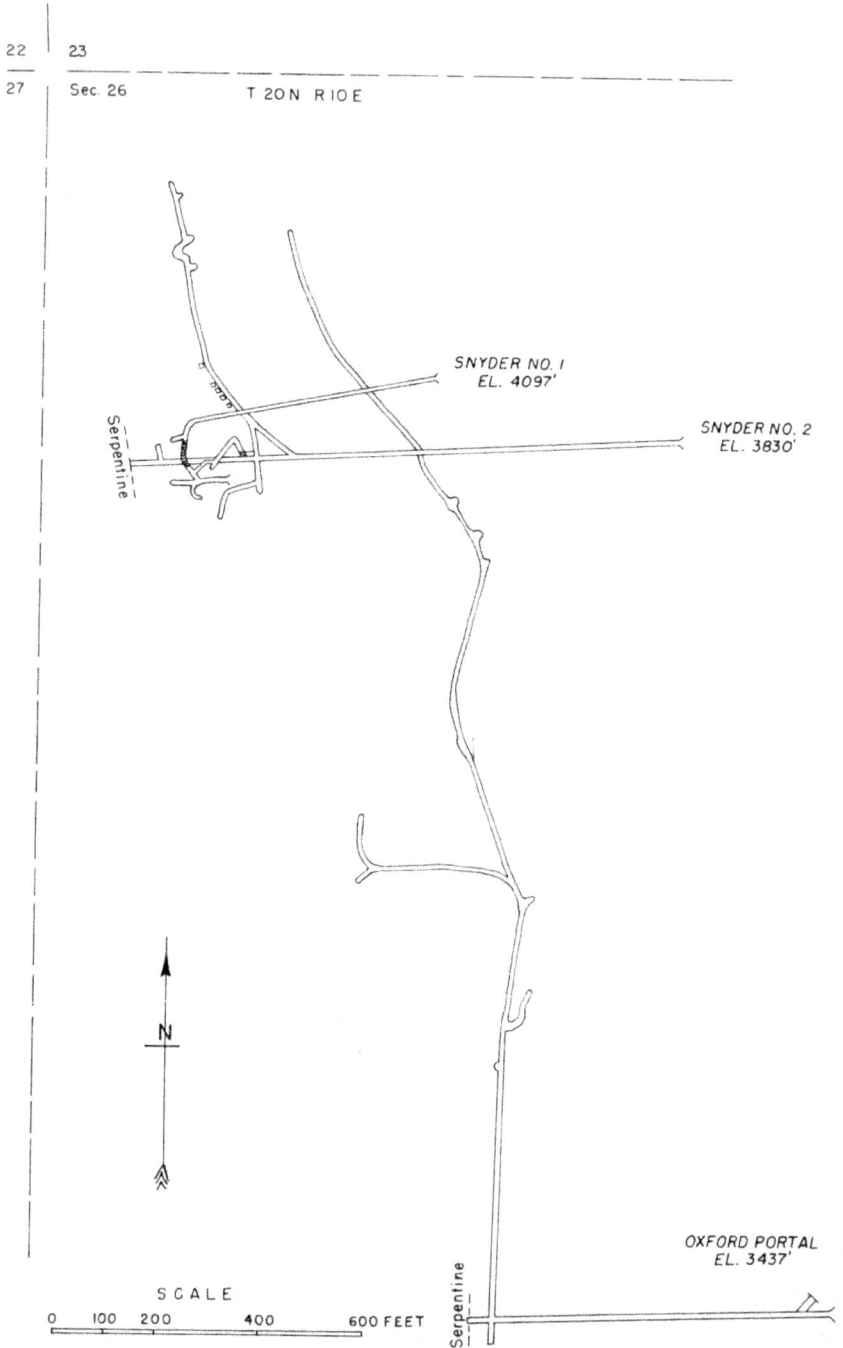

FIGURE 12.   Underground workings of Oxford mine.

below the 350 level of the new winze in the Red Star workings. All veins occur in amphibolite schist of the Tightner formation.

The Osceola vein strikes N. 42° W. and dips 50°-60° northeast for the first 400 feet in from the adit portal and then the strike changes to N. 66° W. and dip to 75°-80° northeast (Ferguson and Gannett, 1932, p. 115). Much of the production from this vein was from near the adit portal where a footwall branch extends to the southeast. A small body of serpentine lies about 100 feet west of this junction. Arsenopyrite is associated with the high-grade ore shoots.

The new vein in the Red Star workings has been the source of all of the production since 1942. It has a north to northwest strike, dips 30 to 45 degrees east, and ranges from 5 to 15 feet in width. The production was largely from stopes located between the 500-foot and 600-foot levels south of the winze.

The mine is developed by a 3,000-foot adit. For the first 1,500 feet beyond the portal, the adit drifts along the Osceola vein in the Osceola claim. The adit then crosscuts 1,200 feet west to intersect the Tightner fault. Three-hundred feet east of the fault a 600-foot inclined winze was sunk to intersect the projected extension of the 16-to-One vein which is believed to extend into this property (Robert McCulloch, personal communication, 1955). The shaft intersected the new or possible extension of the 16-to-One vein on the 500-foot level, and also levels have been run at 350 and 600 feet. The 500-foot level is the most extensively developed level on this new vein. The last work done, which was late in 1954, was on the 600-foot level where the north drift was extended 25 feet, and a winze was started near the 600 station. When in operation, ore was hand-trammed to the winze, hoisted, and then hauled by a storage battery locomotive to the mill.

FIGURE 13.   Red Star mine and mill; camera facing north.

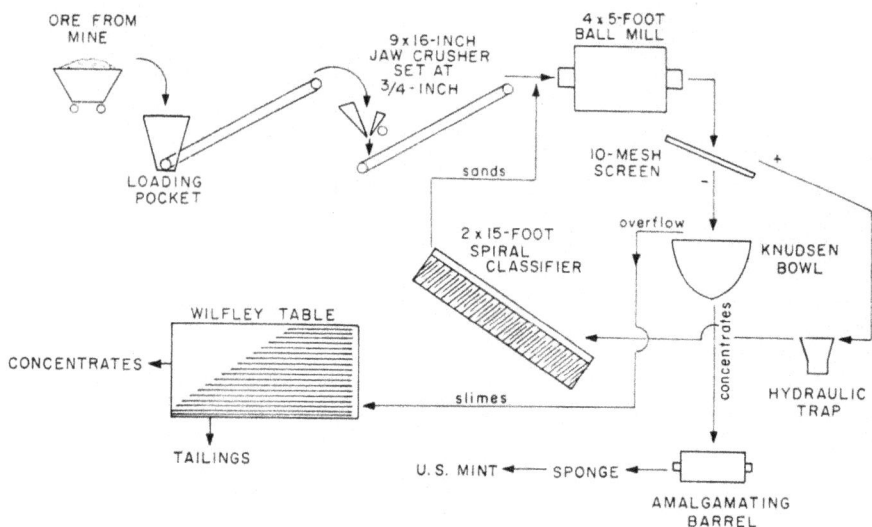

FIGURE 14.   Red Star mill flowsheet.

The mill, which is located immediately northeast of the adit portal, was completed in April, 1953. Ore from the mine was deposited in a loading pocket and then belt-conveyed to the mill. At the mill, the ore was crushed, finely ground, sent to a 10-mesh screen, Knudsen Bowl, hydraulic trap, spiral classifier, and Wilfley table. Sponge gold was sold to the U. S. Mint in San Francisco. Mill capacity is 50 tons; however, when in operation, about 20 tons of ore per day were treated (C. J. Ayres, personal communication, 1955). Seven men worked at the mine and mill.

Sixteen-to-One Mine.   Location: Secs. 3, 4, T. 18N., R. 10E., and sec. 34, T. 19N., R. 10E., M. D. M., in Alleghany. Ownership: Original Sixteen-to-One mine, Inc., 235 Montgomery Street, Room 1611, San Francisco 4, California. C. A. Bennett is superintendent, Willard Van Doren, mine foreman, J. B. Hunley, mill foreman, Wilford Hart, engineer.

The Sixteen-to-One property was located in 1896 and has since acquired claims which were formerly held by the Twenty-one and Tightner mines. The mine, which is the best known and most productive in the Alleghany area, has been continuously active since 1908. It is credited with a total gold production in excess of 730,000 fine ounces (Cooke, 1947, p. 212). Virtually all of the gold production has been from extremely rich ore shoots. Single shoots of ore in the Sixteen-to-One vein have yielded from $200,000 to as much as $1,000,000 in gold (Averill, 1942, p. 44).

The Sixteen-to-One vein occupies a reverse fault on which the amount of movement has been 300 to 900 feet (Ferguson and Gannett, 1932, p. 107). In the southern part of the mine the vein has a general strike of N. 30° W., but in the Tightner workings to the north its strike is essentially north. Although the dip varies greatly, from less than 20

FIGURE 15. Gold Crown mine yard (foreground); Sixteen-to-One mine and mill (background); camera facing east.

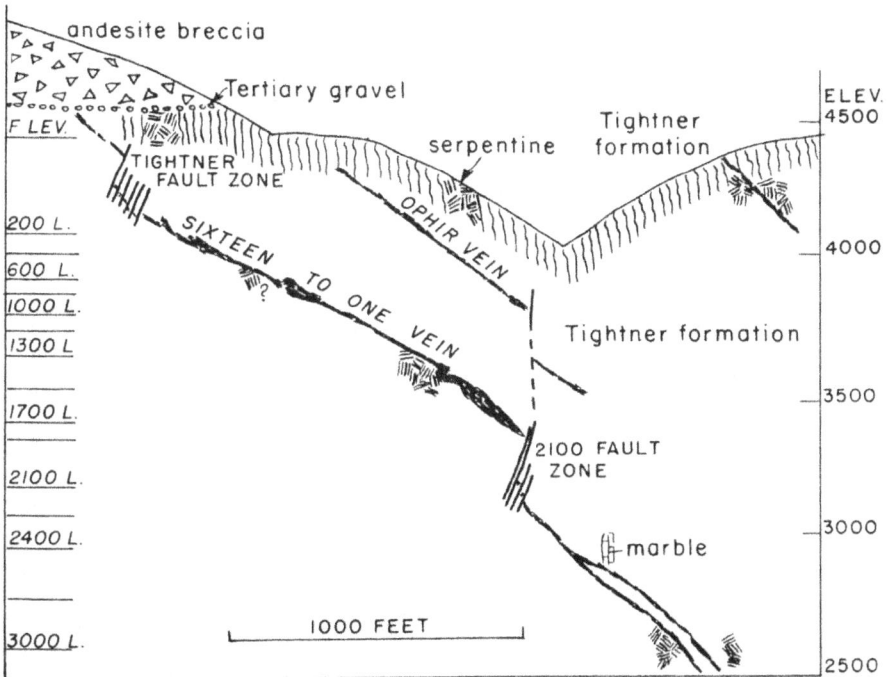

FIGURE 16. Generalized cross-section of Sixteen-to-One vein, Sixteen-to-One mine. (Modified after Cooke, 1947, p. 215.)

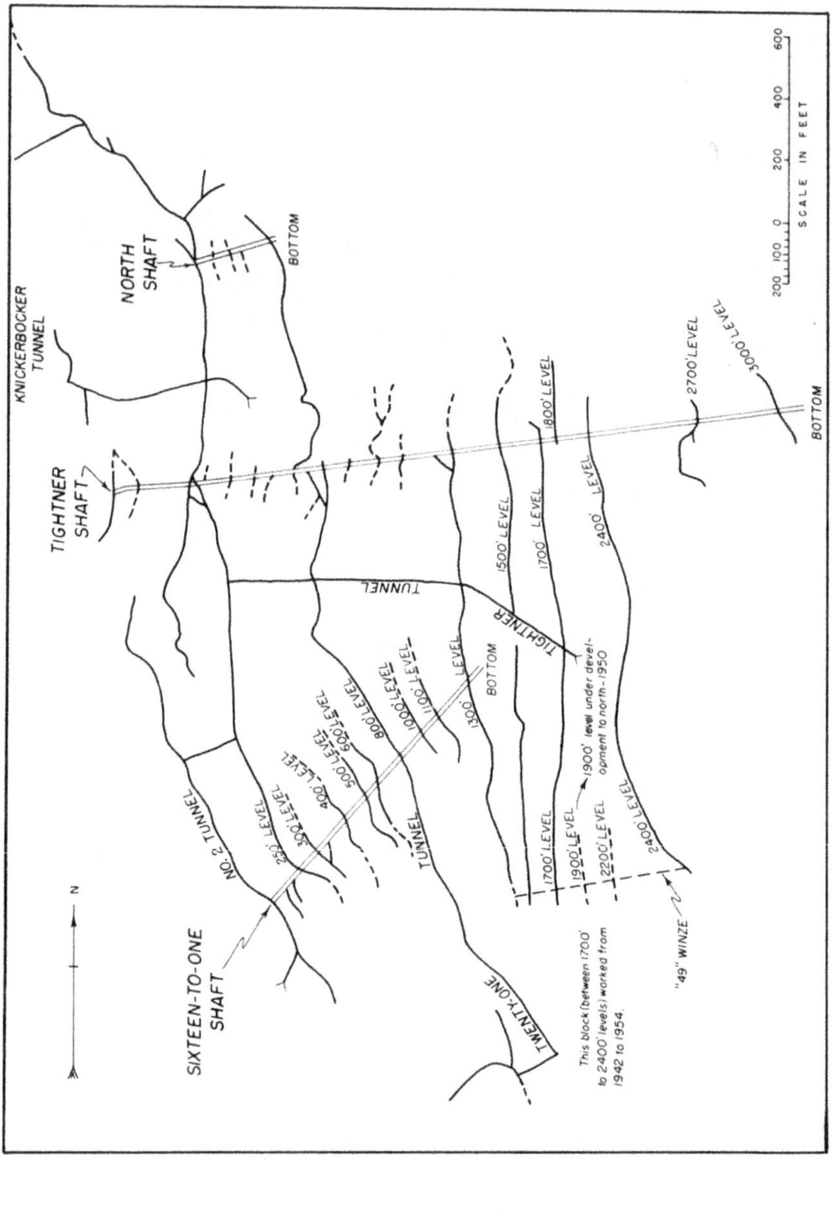

FIGURE 17. Underground workings of Sixteen-to-One mine.

to 60 degrees northeast, the dip of the vein in the most productive parts of the mine ranges between 25 to 30 degrees.

The width of the vein is as much as 50 feet in places, but is usually 5 to 15 feet. The east-dipping Sixteen-to-One vein is cut by two groups of faults, which for the most part dip steeply to the west and have an average strike of N. 24° W. The upper group of faults, known as the Tightner fault zone, cuts the vein in the area between the two shafts between the No. 2 level and the 300-foot level. The lower group of faults, known as the 2100-fault zone, cut the vein between the 1800-foot and 2100-foot levels.

Hornblende schist of the Tightner formation comprises most of the country rock, although rocks of the Kanaka formation are found in the upper levels. Two dikes of serpentine cut the Tightner formation and are faulted by the vein.

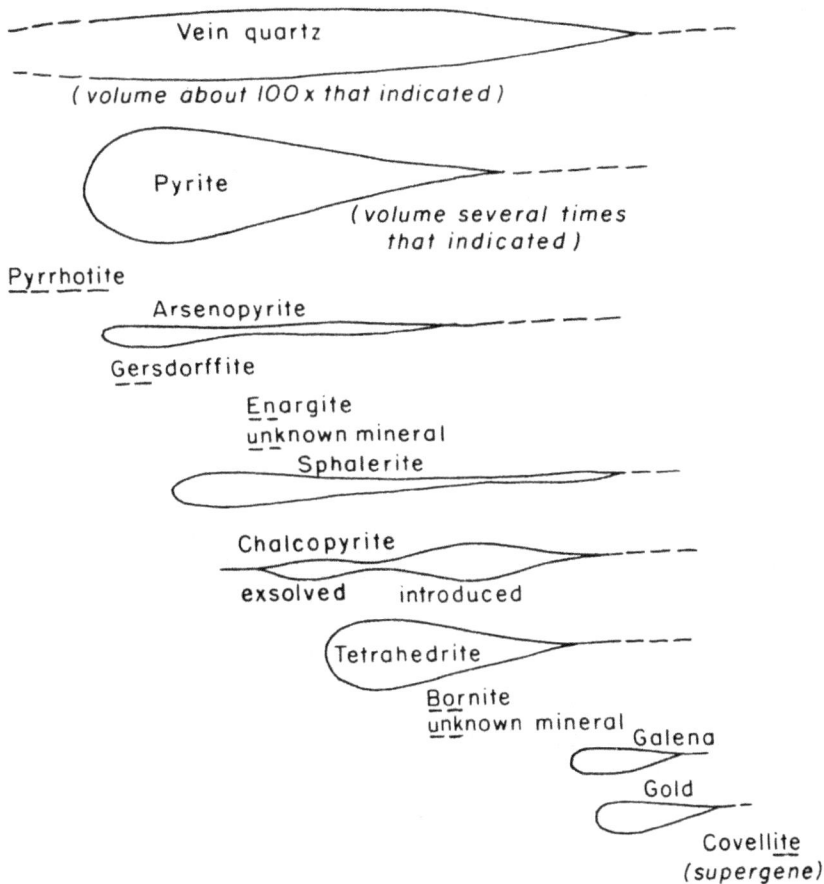

FIGURE 18.   Sequence of mineral deposition in the Sixteen-to-One vein, Sixteen-to-One mine. (After Cooke, 1947, p. 232.)

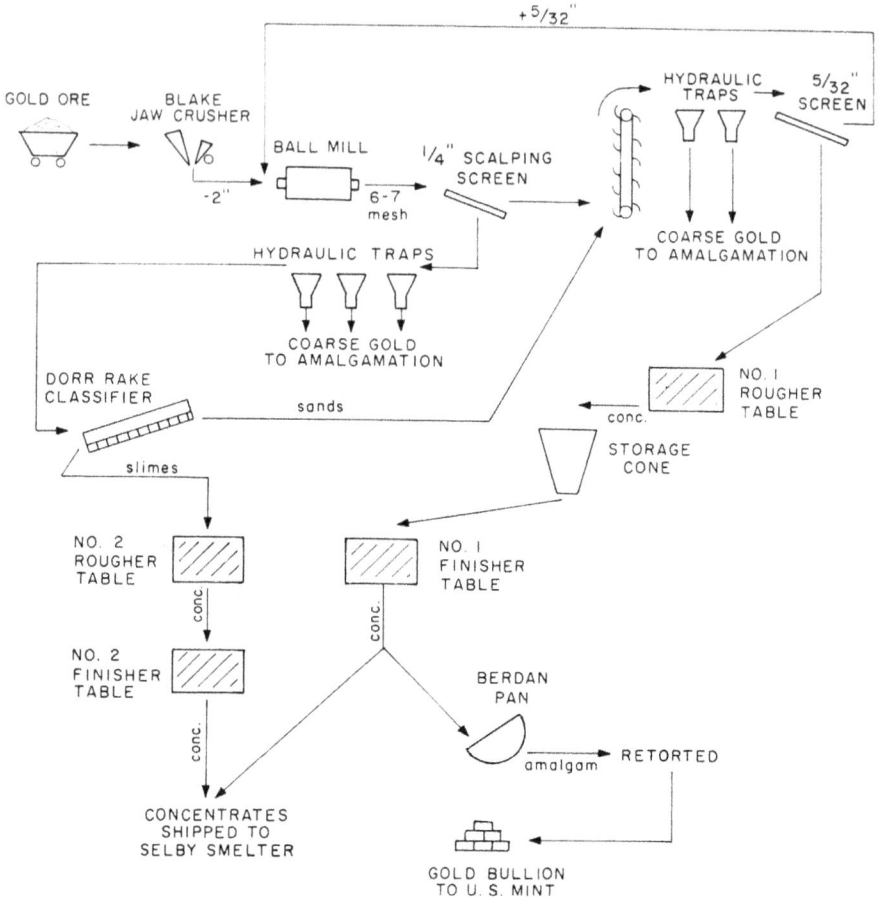

FIGURE 19.   Sixteen-to-One mill flowsheet.

The tenor of the veins is too low for profitable mining and conse-
quently, the veins must be selectively mined so that a higher percentage
of high-grade ore is mined than would be in random mining of the
vein. Since no reliable criterion has yet been found to determine the
position of high-grade shoots in advance of development work, the
shoots are discovered by cutting the more favorable portions of the vein
into small blocks by means of a large amount of development work.

A winze, known as the Sixteen-to-One shaft, is the main working
entry to the mine at the present time. This winze extends from the
No. 2 tunnel down to the 1,300-foot level. From the 1,300-foot level the
49 winze extends down to the 2,400-foot level. Since World War II,
production has been from the vicinity of the 49 winze below the 1,700
level. The lowest level now worked is the 2,400 level. The ore is hoisted
up the 49 winze to the 1,300 level where it is trammed northward to
the Sixteen-to-One shaft. The ore is then hoisted up the Sixteen-to-One
shaft to the No. 2 tunnel and trammed by airtrammer to the mill. The
Tightner shaft collared on the 250 level and extending to the 3,000 level

was at one time the main working shaft for the mine and is still maintained for pumping, servicing and ventilation.

Open timbered stopes utilizing slusher scrapers are typical of the Sixteen-to-One Mine. Mine waste is used for filling old stopes. Drilling is done by Ingersoll-Rand R-58 stopers. In drilling drift and crosscut headings, the stopers are mounted on air bars. Two changes of steel equipped with throw-away bits are used to drill 6-foot holes. Compressed air and electric slushers and two pneumatic mucking machines handle the broken ore. Five storage battery locomotives haul 1-ton side-dump ore cars. A compressed air trammer is used for haulage on the surface.

At the mill the ore is crushed by a primary jaw crusher and secondary ball mill, classified by a Dorr rake classifier, and the fines are concentrated on Deister tables. High-grade ore, which amounts to approximately 50 percent of the total production, is handled and milled separately. The free gold concentrate from the mill, which amounts to approximately 45 percent of the total production, is amalgamated and retorted. The sulfide concentrates are shipped to the Selby smelter.

The mine and mill operate 6 days per week and one and $1\frac{1}{2}$ shifts per day, respectively. Thirty-six men are employed at the mine and mill.

REFERENCES

Averill, C. V., 1942, Mines and mineral resources of Sierra County: California Div. Mines Rept. 38, pp. 7-67.

Averill, C. V., 1949, Sierra County: California Div. Mines Bull. 142, pp. 121-123.

Bowen, O. E., Jr., and Crippen, R. A., Jr., 1948, Geologic maps and notes along Highway 49: California Div. Mines Bull. 141, pp. 78-83.

Cooke, H. R., Jr., 1947, The Original Sixteen-to-One gold quartz vein, Alleghany, California: Econ. Geology, vol. 42, no. 3, pp. 211-250.

Crawford, J. J., 1894, Gold—Sierra County: California Min. Bur. Rept. 12, pp. 260-275.

Crawford, J. J., 1896, Sierra County: California Min. Bur. Rept. 13, pp. 217-385.

Ferguson, H. G., 1915, Lode deposits of the Alleghany district, California: U. S. Geol. Survey Bull. 580, pp. 153-182.

Ferguson, H. G., and Gannett, R. W., 1929, Gold-quartz veins of the Alleghany district, California: Am. Inst. Min. Met. Eng. Tech. Pub. 211.

Ferguson, H. G., and Gannett, R. W., 1932, Gold-quartz veins of the Alleghany district, California: U. S. Geol. Survey Prof. Paper 172, 139 pp.

Goldstone, L. P., 1890, Sierra County: California Min. Bur. Rept. 10, pp. 642-654.

Irelan, William, Jr., 1886, Sierra County: California Min. Bur. Rept. 6, part II, pp. 52-59.

Irelan, William, Jr., 1888, Sierra County: California Min. Bur. Rept. 8, pp. 573-581.

Lindgren, W., 1900, U. S. Geol. Survey Geol. Atlas, Colfax folio (no. 66), 11 pp., 4 maps.

Logan, C. A., 1921, Sierra County: California Min. Bur. Rept. 17, pp. 474-478.

Logan, C. A., 1922, Sierra County: California Min. Bur. Rept. 18, pp. 143, 499-519.

Logan, C. A., 1924, Sierra County: California Min. Bur. Rept. 20, pp. 17-18.

Logan, C. A., 1929, Sierra County: California Min. Bur. Rept. 25, pp. 151-212.

Logan, C. A., 1935, Sierra County: California Div. Mines Rept. 31, pp. 3-4.

MacBoyle, E., 1920, Mines and mineral resources of Sierra County, California: California Min. Bur. Rept. 16, 144 pp.

McKinstry, H. E., and Ohle, E. L., Jr., 1949, Ribbon structure in gold-quartz veins: Econ. Geology, vol. 44, no. 2, pp. 87-109.

O'Brien, J. C., 1951, Sierra County: California Div. Mines Rept. 47, pp. 371-373.

Preston, E. B., 1892, Sierra County: California Min. Bur. Rept. 11, pp. 400-412.

Preston, E. B., 1895, Sierra County gold mill practices: California Min. Bur. Bull. 6, pp. 73-74.

Simkins, W. A., 1923, The Alleghany district of California : Pacific Mining News of Eng. & Min. Journal, vol. 2, pp. 288-291.

Taylor, G. F., 1903, Register of mines and minerals, Sierra County : California Min. Bur., 14 pp., 1 map.

Turner, H. W., 1896, Further contributions to the geology of the Sierra Nevada : U. S. Geol. Survey 17th Ann. Rept., pt. 1, pp. 54-762. (Downieville area, pp. 591-625).

Turner, H. W., 1897, U. S. Geol. Survey Geol. Atlas, Downieville folio (no. 37), 8 pp., 4 maps.

Wiltsee, E. A., 1892, Some additional Sierra County mines : California Min. Bur. Rept. 11, pp. 413-419.

LIST OF LODE GOLD MINES IN THE ALLEGHANY-DOWNIEVILLE AREA THAT HAVE BEEN ACTIVE SINCE 1942

The list of lode gold mines and prospects is arranged alphabetically. The number in the first column refers to the location on the map in pocket. The names and numbers in parentheses in the last column refer to the accompanying bibliography. The first number after the author's name is the year of publication, and is separated from the page reference by a colon. Other references are separated by semicolons. The term "herein" refers to a description in this paper.

LIST OF LODE GOLD MINES IN THE ALLEGHANY-DOWNIEVILLE AREA THAT HAVE BEEN ACTIVE SINCE 1942

No.	Name of claim or mine	Owner (name and address)	Sec.	T.	R.	B & M	Remarks and references
1	Brush Creek	Alpha Hardware & Supply Co., F. F. Cassidy, Nevada City	17, 18, 19	19N	10E	MD	(Lindgren 00:8; Ferguson 15:182; Mac Boyle 20:79; Logan 22:502; Logan 24:17; Logan 29:160; Ferguson and Gannett 32:26, 87–89; Averill 42:21; Bowen and Crippen 48:78; Averill 49:122; O'Brien 51:371; herein).
	Colorado						See Yellow Jacket
2	El Dorado	George S. Fessler Estate, Jane A. Morris, 806 E. Calif. Ave., Sunnyvale	34, 35	19N	10E	MD	Surface clean up and some retimbering of the adit is being done by Roland De Grio. (Preston 92:407; Crawford 96:372; Ferguson 15:174; Mac Boyle 20:85–86; Logan 21:474; Logan 22:503–505; Ferguson and Gannett 32:118–119; Logan 35:3).
3	Eureka	O. P. Bixby et al., 537 W. Grand Ave., Oakland, Calif.	29, 32	19N	10E	MD	A small quantity of gold was produced by Maurice Derrnan in 1946. Leased by C. J. Ayers in 1949. At present, a small amount of rehabilitation work is being done by Mr. Bixby. (Ferguson and Gannett 32:55, 92).
4	Gold Bluff	Mrs. Irene Best, Downieville	23	20N	10E	MD	(Irelan 88:579-580; Crawford 94:266; Crawford 96:376; Mac Boyle 20:110; Logan 24:18; O'Brien 51:371; herein).
5	Gold Crown (Wonder)	Gold Crown Mining Corp., Alleghany	3, 34	18N, 19N	10E, 10E	MD, MD	(Logan 22:519; Ferguson and Gannett 32:102; herein).
	Hub						See Seven Aces
6	Irelan (Ireland)	Yuba-Irelan Mines, Ltd., 1921 Stockton Blvd., Sacramento, Calif.	11	18N	10E	MD	(Lindgren 00:8; Ferguson 15:179; Mac Boyle 20:92; Logan 22:506-507; Logan 24:17; Ferguson and Gannett 32:14, 127-128; herein).
7	Kate Hardy	H. E. Hawn, P.O. Box 748, Grass Valley, Calif.	29, 32	19N	10E	MD	(Ferguson 15:181-182; Mac Boyle 20:95-97; Logan 21:475; Logan 22:507-508; Logan 24:17; Logan 29:166; Ferguson and Gannett 32: 89-91; O'Brien 51:372; herein).

LIST OF LODE GOLD MINES IN THE ALLEGHANY-DOWNIEVILLE AREA THAT HAVE BEEN ACTIVE SINCE 1942—Continued

No.	Name of claim or mine	Owner (name and address)	Location Sec.	T.	R.	B & M	Remarks and references
--	Liberty (Macchaus)	Delbert R. Schiffer, et al., Nevada City.	5	18N	10E	MD	Exploration work and some diamond drilling done between 1946 and 1951 by several lessors (Mac Boyle 20:29; O'Brien 51:372).
--	Macchaus						See Liberty
8	Morning Glory		34	19N	10E	MD	Leased by C. Hawkins who has done a minor amount of rehabilitation work. There has been no recent production. (Mac Boyle 20:105; Logan 22:509-510; Ferguson and Gannett 32:112).
9	Mugwump (Young America)	Mugwump Mining Co., The Vivians, Forest.	28	19N	10E	MD	(Preston 92:407; Crawford 94:275; Crawford 96:384; Mac Boyle 20:53; 133-134; Logan 22:143; Logan 29:194-195; Ferguson and Gannett 32:92-94; O'Brien 51:373; herein).
10	Oriental	Drescher Estate, Financial Bldg., 10th & J Sts., Sacramento and Mrs. W. E. Kliensorge, 2161 Hyde St., San Francisco	33 / 4	19N / 18N	10E / 10E	MD	(Lindgren 00:8; Ferguson 15:176-177; Mac Boyle 20:107-108; Logan 22:511; Logan 24:18; Logan 29:168; Ferguson and Gannett 32:26, 32, 98-101; Logan 35:3; Averill 42:33-35; O'Brien 51:372; herein).
--	Oro	Dr. & Mrs. Phillip Newton, Oakland, Calif.	35 / 2	20 / 19	10 / 10	MD / MD	Some work done in 1946 and 1947 by the Associated Metals Company of Seattle, Washington. (Mac Boyle 20:108; Logan 21:475; Logan 22:737; Averill 49:122).
--	Osceola						See Red Star
11	Oxford	Oxford Quartz Mining Co., Downieville.	26	20N	10E	MD	(Crawford 94:269; Crawford 96:381; Mac Boyle 20:109; Averill 42:35; O'Brien 51:371; herein).
12	Pilgrim (American Hill)		29	19	11	MD	American Hill Mines operated mine from Oct. 15 to Dec. 20, 1946; 600 tons ore yielded 145 oz. gold and 22 oz. silver. Six tons ore shipped to smelter. (Mac Boyle 20:112; Logan 29:168-169.)

LIST OF LODE GOLD MINES IN THE ALLEGHANY-DOWNIEVILLE AREA
THAT HAVE BEEN ACTIVE SINCE 1942—Continued

No.	Name of claim or mine	Owner (name and address)	Location				Remarks and references
			Sec.	T.	R.	B & M	
13	Red Ledge		12	18N	9E	MD	Some recent surface exploration work done by several leasors. There is a 50-foot wide zone of mineralized quartz veinlets in serpentine. (Mac Boyle 20:114-115).
14	Red Star (Osceola, Yellow Jacket)	Tightner Mines Co., 58 Sutter St., San Francisco	34	19N	10E	MD	(Preston 92:407; Ferguson 15:173-174; Mac Boyle 20:108-109; Logan 22,511-512; Ferguson and Gannett 32:78, 79, 114-115, 119-120; O'Brien 51: 373: herein).
15	Roye-Sum	Not determined	4	18N	10E	MD	Old workings were cleaned out and sampled by several leasors; there was no production. (Ferguson and Gannett 32:98).
16	Ruby drift mine	Not determined	15	19N	10E	MD	Intermittent minor lode production from small veins which were encountered in bedrock below the channel. (Preston 93:406; Crawford 94:271-272; Crawford 96:382; Mac Boyle 20:56-57; Averill 42:38-42).
17	Seven Aces (Wyoming, Hub)	W. B. Brinker, et al., 140 Douglas St., San Francisco, Calif.	33	19N	10E	MD	Intermittent development work being done by the owner. (Ferguson and Gannett 32:102; O'Brien 51:372).
18	Seymour	Not determined	4	18N	10E	MD	A minor amount of development was done recently by several leasors.
19	Sixteen to One	Original Sixteen-to-One Mine, Inc., 235 Montgomery St., San Francisco	3, 4 34	18N 19N	10E 10E	MD MD	(Ferguson 15:175; Mac Boyle 20:122-124; Logan 21:475; Logan 22:343, 514-517; Logan 24:17, 18; Logan 29:171-172; Ferguson and Gannett 32:31, 33, 35, 38, 54, 106-110; Logan 35:3; Averill 42:44-47; Bowen and Crippen 48:78; Averill 49:122; O'Brien 51:373; herein).
20	Telegraph	Jack Sheedy, Downieville	4	20N	10E	MD	Small amounts of rehabilitation work done recently. (Preston 93:410-412; Mac Boyle 20:126; Logan 29: 187; Logan 34:3).

LIST OF LODE GOLD MINES IN THE ALLEGHANY-DOWNIEVILLE AREA THAT HAVE BEEN ACTIVE SINCE 1942—Continued

No.	Name of claim or mine	Owner (name and address)	Location Sec.	Location T.	Location R.	Location B & M	Remarks and references
--	Tightner						Part of the Sixteen-To-One
							(Ferguson 15:171-173; Mac Boyle 20: 127-129; Logan 21: 475; Logan 22:143, 517-519; Logan 24:17, 83; Ferguson and Gannett 32:106-109; Bowen and Crippen 48:78; herein under 16-to-1 Mine).
--	Twenty-One						Part of the Sixteen-to-One Mine
							(Ferguson 15:174; Mac Boyle 20:130-131; Bowen and Crippen 48:78; herein under 16-to-1 Mine).
--	Wonder						See Gold Crown
--	Wyoming						See Seven Aces
21	Yellow Jacket (Colorado)	Yellow Jacket Cons. Gold Mines, Ltd., 120 Chester Ave., Bakersfield, Calif.	34	19N	10E	MD	Some development work done 1947. During 1952-1953, the Jolly Jack Company partially reopened the main adit. (Mac Boyle 20:81; Logan 22:511-512; Ferguson and Gannett 32:119-120).

HISTORY OF BORAX PRODUCTION IN THE UNITED STATES

BY W. E. VER PLANCK *

The American borax industry is very nearly synonymous with the California borax industry. In the latter part of the 19th century, however, borax production in Nevada equalled that of California and even exceeded it at times. Oregon also produced borax at about the same period.

Borax was one of the first of the nonmetallic mineral commodities to have been considered worthy of the attention of the prospector, the engineer, and the promoter. Like the precious metals, borax has held forth the enticing promise of fortunes to be made with little effort. To the popular mind, it is irrevocably associated with Death Valley and a host of semi-legendary episodes. While the borax industry can look back on a past that is not without its dramatic moments, today it is a mature industry in which profits are made only by hard work coupled with the best of engineering and business practices.

A century ago, borax was a high-priced commodity imported from England, mainly for pharmaceutical purposes. Consumption amounted to but a few hundred tons a year. The principal source of raw materials was the sassolite deposits of Italy, which were developed in the early 18th century. Prior to that time a very impure borax called tincal was obtained from the interior of Asia. Refineries were established first in Venice and later in other European cities and in England.

Lake County Period (1864-72). Borax was first detected in California in January, 1856, by Dr. J. A. Veatch while making an analysis of the water of Tuscan Springs in Tehama County. The search for a workable deposit culminated in September of the same year in the discovery of the small saline lake near Clear Lake in Lake County that has since been called Borax Lake. This lake, which occupies an isolated basin in an area of Quaternary volcanism, contains a strong brine, rich in sodium chloride, carbonate, and borate. In addition, the bottom mud was found to be crowded with borax crystals ranging up to several inches in length. Some time passed before the deposit could be studied and a company formed to work it, and it was not until 1864 that the California Borax Company with W. O. Ayers as Superintendent, actually produced borax.

The methods used by the California Borax Company were simple. Iron coffer dams, 4 feet square by 6 feet deep were floated out by means of a raft and dropped into the crystal-bearing mud. Then the mud within the coffer dam was shoveled out and the borax crystals were washed from it. In this way, and in spite of the loss of the small crystals, as much as 900 pounds of borax were obtained from the mud within a single coffer dam. The crystals recovered were refined by dissolving in hot water and recrystallization. From 1864 to 1868 an average of 300 tons a year of refined borax was produced, which sold for 28 to 35 cents per pound. This material according to an analysis made in 1865 was 99.95 percent pure.

Before long the company felt that the supply of crystals that could be recovered in this way was limited, and means of increasing the

---

* Geologist, California State Division of Mines.

MINES AND MINERAL RESOURCES OF EL DORADO COUNTY, CALIFORNIA

By William B. Clark * and Denton W. Carlson **

OUTLINE OF REPORT

* Associate mining geologist, California Division of Mines.
** Terminal superintendent, Tidewater Associated Oil Company; formerly assistant mining geologist, California Division of Mines.

Manuscript submitted for publication May 1, 1956.

ABSTRACT

El Dorado County lies within the northern portion of the Sierra Nevada, chiefly on the west slope. Coloma, on the American River, was the site of James C. Marshall's gold discovery of 1848. Soon afterward thousands of gold seekers arrived to work the streams. Mining, lumbering, agriculture, and vacation and resort services are the major industries of the county.

Rocks exposed in El Dorado County range from Paleozoic to Recent in age. The entire county is underlain by the Bedrock Series of the Sierra Nevada. This series consists of steeply dipping, intensely folded and faulted metamorphic rocks of Paleozoic and Mesozoic age that have been invaded by several types of igneous rocks, chiefly granitic. Associated with the intrusion of the igneous rocks was the formation of gold, silver, copper, and chrome deposits. Over the bedrock are nearly flat-lying gold-bearing gravels and volcanic and sedimentary rocks of Tertiary age. Glacial moraines of Pleistocene age cover large portions of the higher Sierra in eastern El Dorado County. Lake Tahoe was formed by faulting during the Tertiary time.

From 1880 to 1953, the total mineral output of the county was $57,122,787. Gold has been the most important single mineral commodity produced, followed by limestone and limestone products. The most productive gold mines are on the Mother Lode belt. Gold also has been produced from the East and West gold belts. The Mother Lode gold belt is a system of northwest-trending gold-bearing quartz veins that traverses the central portion of the county. The seam gold deposits in the north end of the county have been worked by both placer and hardrock methods. Major Mother Lode gold mines in the county include the Union, Pacific, Beebe, Sliger, and Church mines. Seam deposit at Georgia Slide were highly productive.

East Belt gold veins are smaller but usually richer than those of the Mother Lode. Noted East Belt gold mines include the Mount Pleasant, Alhambra, and the Hazel Creek, the last being the principal source of gold in the county. West Belt deposits are low-grade but usually quite large. Among the more productive West Belt mines were the Big Canyon, Zantgraf, Crystal, and Shaw.

Substantial quantities of gold were produced from Eocene channel deposits in hydraulic and draft mines, particularly in the Placerville area. Other channels were worked east of Georgetown, Grizzly Flat, and Indian Diggings. During the 1930's, many Recent stream gravels were worked by dragline dredges.

There are a number of extensive deposits of crystalline limestone in the county. At present, the El Dorado Limestone Company and the California Rock and Gravel Company produce limestone for use in the beet sugar industry. These concerns also ship limestone to the Diamond Springs Lime Company, for use in the manufacture of chemical and metallurgical lime. Lime also is produced at the Semon Lime Company plant near Rattlesnake Bridge. For many years limestone was produced from the Cool-Cave Valley deposit for use in cement manufacturing.

Large quantities of chromite have been produced at the Pillikin mine, which was active until 1955. Also active until 1955 were several custom chrome mills. Chromite also has been mined from deposits in the Latrobe and Georgetown areas.

Copper has been produced from the foothill copper belt and as a by-product of gold mining. At present, copper ore from Amador County is concentrated at the Volo mill near Placerville. Silver, lead, and zinc have been produced as by-products from gold and copper mining.

Tungsten prospects recently have been found east of the Mother Lode. At present a prospect near Grizzly Flat is under development. Other metals found in El Dorado County include iron, manganese, molybdenum, nickel, platinum, and quicksilver.

Slate is mined at Chili Bar. It is fine-ground and marketed as roofing material. Dimension slate was produced from several nearby properties many years ago. Soapstone is mined near Latrobe, milled in the San Francisco Bay area, and marketed as an insecticide carrier. Rhyolite tuff is quarried east of Placerville and used as ornamental stone in home construction.

Sand and gravel are mined near Placerville. Crushed rock is produced south of Placerville for use as asphalt mix. Broken granodiorite has been used as rip-rap on various roads in the county. Sand and limestone are mined for use as road metal.

INTRODUCTION

El Dorado County is in the north portion of the Sierra Nevada. It is bounded on the north by Placer County, on the west by Sacramento County, on the south by Alpine and Amador Counties, and on the east by part of Alpine County and the State of Nevada. In 1850, it was created as one of California's original 27 counties. El Dorado County was named from the Spanish phrase meaning "The Gilded One."

History. Coloma, the scene of Marshall's historical gold discovery, is on the South Fork of the American River 8 miles northwest of Placerville. In 1847, John Sutter, the grantee and almost feudal baron of a large Mexican land grant at what is now Sacramento, signed a contract with James Marshall to erect and operate a sawmill on the American River. The mill was under construction late that year. It then was decided to enlarge the tailrace to allow more water to run off. On the morning of January 24, 1848, while inspecting the tailrace, Marshall noted a yellow flake in shallow water and picked it up. After testing it for malleability, other flakes were obtained. These were taken to Sutter's Fort where more tests were applied. The tests proved beyond a doubt that gold had been discovered.

Sutter and Marshall tried to keep the discovery a secret. However, word leaked out from several sources, and soon it was known in Benicia and San Francisco. The first published announcement of the discovery of gold appeared in the San Francisco newspaper *The Californian* on March 15, 1848. Among the first miners to arrive at the scene of discovery were Mexicans from Sonora who previously had worked placers in Los Angeles County. Soon afterward, thousands of gold seekers arrived in El Dorado County and spread out in all direction in the Sierra

Nevada. Shallow placers in nearly every stream in the western part of the county were worked at this time. Lode mining began in El Dorado County in 1851 at Nashville.

Towns such as Nashville, Spanish Flat, Grizzly Flat, Chili Bar, Fairplay, and Diamond Springs sprang up. Placerville, which was variously known as Old Dry Diggins and Hangtown, was incorporated in 1854. In 1850, El Dorado County was founded as one of the original 27 counties in California. During the 1850's, it was one of the most populous counties in the state. In 1855, the Sacramento Valley Railroad, the first rail line in California and now part of the Southern Pacific system, was built to connect Sacramento and Shingle Springs. Later it was extended to Placerville. During the Comstock Lode boom of the 1860's and 1870's, the county prospered from the Placerville wagon and stage road, which followed along what is now U. S. Highway 50. The road was one of the main routes from San Francisco to Virginia City. Another well-traveled route of those days was the wagon and stage road, which ran east along the Rubicon River via Georgetown, Wentworth Springs, and Rubicon Springs, and terminated at Lake Tahoe.

With the decline of the Comstock mines, the economy of El Dorado County became more diversified. Lumbering, fruit growing, and stock raising became important. Later, with the improvements made in transportation facilities, vacation and resort services increased greatly, especially in the Lake Tahoe area. Interest in the numerous landmarks including the preserved and restored buildings of the old gold mining camps also has accounted for an increase of tourists in the county.

Industries. Lumbering is the largest industry in El Dorado County. According to the El Dorado County Chamber of Commerce, approximately $16,000,000 worth of lumber and lumber products were produced in the county in 1952. Resort and vacation services, the second largest source of income in the county, accounted for more than $7,500,-000 in 1952, while agricultural production totaled $4,069,000. The main agricultural products are fruit, livestock, and poultry. Placerville is the chief agricultural center of the county. In 1953, the mineral output of the county totaled $1,992,069.

The principal lumber-producing areas are in the central and east-central portions of the county. Large sawmills are at Camino, Diamond Springs, Georgetown, Omo Ranch, Pacific House, and Shingle Springs, and there are several logging field camps that are moved from time to time. Until recently two of the major concerns were served by narrow-gauge logging railroads. One extended from Camino northeast to Pino Grande and the other southeast from Diamond Springs to Caldor.

For many years the Lake Tahoe region has been a popular resort area. Noted resorts are at Stateline, Bijou, Al Tahoe, Meeks Bay, and Camp Richardson. Echo and Fallen Leaf Lakes also are popular vacation areas.

Geography. El Dorado County lies within the Sierra Nevada, most of it being on the western slope. It has an area of 1,725 square miles, 44 percent of which was owned by the Federal Government in 1950. Elevations range from 350 feet above sea level in the western foothills to 10,881 feet at Freel Peak in the extreme eastern portion of the county. Other prominent peaks in the county which are along the eastern Sierran crest are Job's Sister (10,823 feet), Monument Peak (10,085 feet), Pyramid Peak (9,983 feet), and Mount Tallac (9,725 feet).

The principal streams in the county are the west-flowing American, Rubicon, and Cosumnes Rivers, all of which have cut deep precipitous canyons. One-fourth of Lake Tahoe, the largest lake in the Sierra Nevada, is in the east portion of El Dorado County. Other large lakes in the county are Fallen Leaf, Echo, Aloha, and Loon.

Placerville, which had a population of 3,726 in 1950, is the county seat and the largest town. Other towns in the county include Camino, Coloma, Diamond Springs, El Dorado, Georgetown, Shingle Springs, Pollock Pines, Al Tahoe, Tahoe Valley, and Bijou.

U. S. Highway 50 traverses the central portion of the county in an easterly direction. Also the county is served by State Highway 49, which trends north via Placerville; State Highway 89, which runs north along the west border of Lake Tahoe; and State Highway 88, which runs east along the southern border of the county from Cook's Station to Peddler Hill. There are a number of county roads, many of which are paved. Also there are a number of privately owned logging roads.

The county is served by the Southern Pacific Railroad, which runs via Latrobe, Shingle Springs, Diamond Springs, and terminates at Placerville. A lumber railroad runs from Placerville to Camino.

GEOLOGY

El Dorado County is underlain predominantly by crystalline rocks of Mesozoic and Paleozoic age. Metasediments of the Paleozoic Calaveras group underlie nearly a third of the county, and granitic rocks roughly another third. Jurassic metasediments and metavolcanics and Tertiary volcanic and detrital rocks underlie most of the remaining area. The strongly folded Paleozoic and Mesozoic metamorphic rocks, together with the granitic rocks and serpentine intruding them, have been conveniently grouped together under the term "Bedrock series." The Tertiary rocks, which have been laid down upon the upturned edges of the older metamorphic rocks, are commonly grouped together under the term "Superjacent series." The Superjacent series is generally flat-lying or else very gently dipping in contrast to the steeply dipping Bedrock series.

Structure

Structure of the metamorphic and intrusive igneous rocks making up the Sierran bedrock complex is not always clear. Structure sections accompanying the U. S. Geological Survey folios for El Dorado County show a series of beds of metamorphic rocks with continuous easterly dips broken only by intrusions. However, the metamorphic beds actually are a complicated series of folds that have been heavily faulted. By separating the Amador group, Taliaferro has been able to work out the intricate structure of the bedrock west of the Mother Lode in southwestern El Dorado County along the Cosumnes River (Taliaferro, 1943, pp. 285, 306). The repetition of beds of Mariposa slate and greenstone, especially north of Placerville, are results of this intricate folding.

The Mother Lode vein system was formed in fissures that developed within a zone of reverse faulting. The zone trends north and in El

FIGURE 1. Greenstone showing schistose structure, in roadcut north of Placerville.

Dorado County is in Mariposa slate. In most areas the displacement is unknown, but to the south in the upper part of the Argonaut mine in Amador County, the measured displacement of beds along the fault fissure is 120 feet (Knopf, 1929, p. 67). At Garden Valley, the Mother Lode system splits into two branches, one extending north through Georgetown and the other northwest through Greenwood.

The ultrabasic intrusions, which now are largely serpentinized, trend north-northwest and reflect the major structural trend of the region. The acidic batholithic masses were intruded after the beds had been folded (Taliaferro, 1943, p. 285).

Major faulting along the east flank of the Sierra Nevada has had a profound influence on the later geologic history of the county. Faulting occurred on the east flank during the end of the Miocene and at the beginning of the Pliocene epoch, and the Sierra Nevada was elevated and became asymmetrical in form with a broad western slope and a short eastern slope (Hinds, 1952, p. 18). Between this fault zone and the Carson Range in Nevada a section sank, forming the basin now partly occupied by Lake Tahoe. During the end of the Pliocene and beginning of the Pleistocene this deformation was renewed on a major scale.

Rock Units

Calaveras Group (Carboniferous to Permian). The Calaveras group is an undifferentiated suite of metamorphic rocks ranging chiefly from Carboniferous to Permian in age—the oldest known rocks in the county.

It is found in two general areas, one east and one west of the Mother Lode. East of the Mother Lode, it consists of a succession of beds of black clay slate, metamorphosed sandstones, quartzite, and mica schist with scattered lenses of crystalline limestone (Lindgren and Turner, 1894, p. 3). West of the Mother Lode the succession consists of highly compressed black slate and sandstone with some chert and meta-chert, and several limestone lenses. Coral and crinoid debris is present in portions of the limestone. Several small tactite bodies have been developed in rocks of the Calaveras group.

Undifferentiated Metamorphic Rocks (Triassic to Jurassic). West and southwest of Lake Tahoe on both sides of the Sierran crest are detached masses of slate, schist, quartzite, and greenstone. Because of their position in what is possibly a continuation of the Sailor Canyon formation, which lies to the north, these rocks have been classified as Triassic to Jurassic in age (Lindgren, 1896, p. 5).

Greenstone and Metasediments (Partly Upper Jurassic). This group occupies an extensive part of the west portion of El Dorado County. It includes rocks which have been mapped as amphibolite, metadiabase, and hornblende porphyrite, as well as rocks of the Amador group of Upper Jurassic age (Taliaferro, 1943, p. 284). The type section of the northern part of the Amador group crops out along the Cosumnes River in southwestern El Dorado County. It is divided into two formations, the Cosumnes below and Logtown Ridge above. The Cosumnes formation consists of sheared sandstone, slate, and a thick basal conglomerate. The Logtown Ridge formation is composed of massive porphyritic augite andesite and coarse andesitic agglomerate.

Mariposa Formation (Upper Jurassic). The Mariposa formation crops out as a north-trending belt that passes through Placerville, and as a northwest-trending belt along the west border of the county. The formation consists of dark clay slate in uniform and nearly vertical beds. Small cross-bedded sandstone lenses are found locally in the slate.

Auriferous Gravels (Tertiary). The most extensive gold-bearing gravel deposits of Tertiary age in the county are those of the Tertiary South Fork of the American River in the Placerville area. The gravel is composed of pebbles and boulders of quartz, chert, granitic and volcanic rocks interbedded with clay and sand. Considerable amounts of placer gold are present near or at the base of these deposits.

Valley Springs Formation (Middle (?) Miocene). Flat or nearly flat beds of the Valley Springs formation are found in and east of Placerville. They consist predominantly of light to buff-colored rhyolite tuff. The tuff usually is fine grained and contains small crystals of black biotite. Smaller amounts of breccia, conglomerate, and siltstone are present.

Mehrten Formation (Upper (?) Miocene and Pliocene (?)). The Mehrten formation, occupying many of the interstream ridges in the east-central portion of the county, consists chiefly of volcanic debris composed of boulders, cobbles, pebbles, and, in some places, angular

FIGURE 2. Slate of the Mariposa formation exposed in a roadcut near Kelsey.

fragments of porphyritic andesite mingled with volcanic ash and some sandy detritus of non-volcanic origin. A few flows of massive andesite are exposed in the eastern portion of the county.

Basalt (Pleistocene). Several lava flows of early Pleistocene age cap some of the ridges in eastern El Dorado County. They are composed of dark fine-grained olivine basalt.

Glacial Moraines (Pleistocene). Extensive moraines deposited by Pleistocene glaciers are west and south of Lake Tahoe. They are composed of exceedingly rough angular boulders of all sizes mixed with sand and gravel and finer detritus.

Alluvium (Quaternary). Sand and gravel deposited in lake beds, and sand, silt, and gravel in and adjacent to present stream channels make up the Quaternary alluvial deposits.

Intrusive Rocks

Serpentine and Associated Rocks. In the western portion of the county are a number of north- to northwest-trending serpentine bodies. Present in them are smaller amounts of pyroxenite, peridotite, and dunite, and it is from these rocks the serpentine is derived. Considerable amounts of gabbro, which is associated with serpentine are present in some areas (Cater and others, 1951, p. 115). These rocks were intruded prior to the emplacement of the granitic rocks.

Granitic Rocks. Extensive bodies of granitic rock occupy much of the eastern portion as well as several areas in the western portion of the county. Granodiorite is the predominant rock type. However, granite, diorite, and quartz porphyry as well as rocks that have been classified as gabbrodiorite by Lindgren and Turner (Lindgren and Turner, 1894, pl. 2) are included in this group.

FIGURE 3.   Cascade Lake and Mount Tallac, Lake Tahoe area. Camera facing south.

Geologic History

During late Paleozoic time, El Dorado County was covered by a vast open sea. Large amounts of sediments that later gave rise to the various metamorphic rocks of the Calaveras group were deposited. These included mud, sand, calcareous ooze, and chemically deposited silica containing manganese. Toward the end of the Paleozoic era, a crustal disturbance destroyed part of the sea basin, partially folded the sedimentary rocks, and created a land area of unknown extent.

The sea again advanced over what is now the Sierra Nevada. Toward the end of the Jurassic period sedimentary and volcanic rocks including pillow basalts and breccias were deposited at the bottom of this sea. Part of this series is commonly called the Amador group. Marine deposition of fine-grained sediments of the Mariposa formation followed.

In late Jurassic time, orogenic processes completely changed the character of the Sierra Nevada. There was an almost complete withdrawal of the sea. The already folded Paleozoic rocks and nearly flat-lying

FIGURE 4.   High Sierra topography, shaped by glacial action. Lovers' Leap and canyon of the upper South Fork of American River. Camera facing west.

Mesozoic rocks were intensely folded and faulted into a series of complicated northwest-trending folds, and the rocks themselves were metamorphosed. Fine-grained sedimentary rocks were changed to slate, siliceous sediments to quartzites and metacherts, volcanic rocks to amphibolite and chlorite schist and greenstone, and calcareous sediments to crystalline limestone.

Before the complete destruction of the sea basin, the area was intruded by ultrabasic rocks. Chromite, which was present in the ultrabasic rocks, was concentrated by magmatic segregation. Most of these rocks subsequently were altered to serpentine. Soon afterward a sequence of granitic rocks was emplaced on a major scale beginning with the more basic types and followed by the more acidic varieties. Copper and zinc were deposited in what is now the lower Sierran foothills. Gold and silver with some copper and lead were deposited in quartz veins deep within the folded crust. Some tungsten and copper deposits were formed in contact metamorphic zones near the borders of the granitic bodies.

A very long period of erosion followed during Cretaceous time. The rocks lying above the gold deposits were stripped away, and the gold-bearing veins themselves were cut deeply. The elevation of the old Sierra Nevada was greatly reduced, and broad river valleys developed. Near the end of the Cretaceous period, the sea advanced again to the west margin of the ancestral Sierra Nevada. Thin marine deposits of Cretaceous age lie just to the west of El Dorado County.

By Eocene time, the Sierra Nevada had been worn down greatly. The climate at this time was subtropical, and there was much chemical decay of the bedrock. Deposits of clay, quartz, sand, and lignite were formed along the margins of the Eocene sea west and south of El Dorado County. Gold and other weather-resistant material, such as quartz pebbles from the veins, were concentrated in streams as placer deposits. Some of these Eocene placers were unusually rich in gold.

Volcanic activity began in the Sierra Nevada at or near the close of the Eocene epoch. Rhyolite ash fell in the lower elevations, while higher up in the mountains both flows and ash falls were deposited. The stream channels were choked and the drainage system completely changed. The new stream channels that developed were characterized by deposits containing abundant rhyolite pebbles. These deposits are known as the "inter-rhyolite gravels." After the close of the emission of rhyolite, the volcanoes began to emit andesitic lavas. Only near the summit of the range are massive andesite flows found, the greater part of the Sierran slope being covered by extensive mud flows, the first containing sands and clays, the later containing great quantities of andesite fragments. The streams again were forced into new channels, depositing what is known as the "interandesitic gravels." These intervolcanic channels commonly were leaner in gold than the Eocene channels. Volcanic activity reached its climax in late Miocene or early Pliocene times.

The basin now partly occupied by Lake Tahoe developed as a depressed fault block by dislocations occurring during Miocene-Pliocene times. Two principal faults, one east and one west of the lake evolved. Several times in the past Lake Tahoe was of greater size and depth than it is now.

In late Pliocene or early Pleistocene time, the Sierra Nevada was re-elevated on a major scale. This was accomplished by faulting along the east flank. The west-flowing rivers and streams cut deep canyons in the newly uplifted area and removed much of the volcanic cover. The Tertiary channel deposits were exposed and subjected to erosion. Stripping of the gold lodes continued. Glaciers appeared in the higher elevations of the Sierra during the Pleistocene epoch. In El Dorado County they occupied extensive areas west and south of Lake Tahoe and elaborately carved the topography of this region. Echo, Fallen Leaf, and Loon Lakes were formed by glacial action. The glaciers retreated in early Recent time, and the present climate and topographic features developed.

MINES AND MINERAL RESOURCES
Metallic Minerals

Chromite

El Dorado County contains the largest known chromite deposits in the Sierra Nevada. The total recorded production of chromite for the county is about 40,000 tons. Estimated reserves are about 600,000 tons of material containing at least 5 percent chromic oxide (Cater and others, 1951, p. 119). The largest deposits are at the Pillikin mine in the Flagstaff Hill area in the western portion of the county. The bulk of the total chromite production of the county has been from this mine. Other chromite-bearing areas are in and near serpentine bodies in the Latrobe, Clarksville, Webber Creek, Pilot Hill, Garden Valley,

Summary of economic geology of El Dorado County.

Geologic age		Rock units	Rock types	Mineral deposits
Quaternary	Recent	Alluvium	Silt, sand, gravel	Placer gold, platinum, sand and gravel
Tertiary	Pliocene	Mehrten	Andesitic detritus	Building stone
	Miocene	Valley Springs	Rhyolite tuff	Building stone, aggregate, manganese
	Eocene	Auriferous gravel	Gravel, some clay and sand	Placer gold, platinum, gravel, quartz, clay
Mesozoic		Veins	Vein quartz	Gold, silver, copper, lead, zinc, quartz
		Tactite (silicated limestone)	Tactite	Tungsten, copper, gold, silver
	Jurassic	Granitic rock	Granodiorite, granite, diorite, gabbro, etc.	Riprap, road metal, sand, building stone
		Serpentine	Serpentinized peridotite, pyroxenite, gabbro, etc.	Chromite, asbestos, road metal, soapstone
		Mariposa	Slate	Slate
		Greenstone and metasediments	Metavolcanic rocks, some metasediments	Building stone, aggregate
Paleozoic	Permian Carboniferous	Calaveras group	Metasandstone, slate, quartzite, mica schist, metachert, crystalline limestone	Limestone, marble, manganese, slate

Table 1. Mineral production of El Dorado County, 1880-1953.

Year	Gold value	Silver value	Copper Pounds	Copper Value	Lime Tons	Lime Value	Limestone Tons	Limestone Value	Slate Squares	Slate Value	Misc. stone Value	Misc. stone Amount	Misc. and unapportioned Value	Substance
1880	$389,383	$208					1							
1881	550,000	900												
1882	600,000													
1883	530,000													
1884	575,000	16,000												
1885	35,000													
1886	619,992	1,822												
1887	706,871	365												
1888	650,000	500												
1889	427,638	408												
1890	204,583	275												
1891	173,279	359												
1892	198,321													
1893	244,610	1,220			1,600	$8,000			1,800	$11,700				
1894	366,707	355			4,560	28,500			1,350	9,450				
1895	700,101	418			706	4,158			500	2,500				
1896	812,289	534			2,160	6,750	500		400	2,800				
1897	674,625	886			538	3,330		$250	400	2,800				
1898	501,966	4,174			1,270	7,335			600	4,500				
1899	401,497	8,414	3,125	$500	1,200	6,000			3,500	25,250				
1900	368,541	25,129			1,760	11,000			5,100	38,250			$251,820	Unapportioned, 1900-09.
1901	232,036	5,977			3,936	16,176			4,000	30,000				
1902	335,031	52	2,128	319	896	7,000								
1903	277,304				2,058	7,075								
1904	474,994				1,482	6,946			6,000	50,000				
1905	384,735	2,525	160,000	24,960	3,075	21,138	1,051	5,775	4,000	40,000		10 tons	162	Asbestos.
1906	431,746	2,690			1,782	16,193			10,000	100,900		112 tons	2,625	Asbestos.
1907	319,177	2,301			2,517	20,192	5,384	15,318	7,000	60,000		20 tons	1,030	Asbestos.
1908	312,033	5,504	603	122	2,212	14,541			6,000	50,000	$1,600	200 M	8,000	Paving blocks.
1909	238,384	1,299		83	1,808	9,944			6,961	45,680	530	3,763 tons	5,645	Sand (glass).
1910	171,304	967			2,414	12,309			1,000	8,000	2,616	1,200 tons	1,800	Sand (glass).
1911	133,967	1,010	696		2,244	11,218	1,000	1,000			5,465	3,701 lbs.	167	Lead.
1912	105,565	843		107							4,375			
1913	62,688	250			2,210	12,082					1,654		4	Lead.
1914	133,886	654	417	73	2,546	12,872					2,600	90 lbs.	5,250	Slate and soapstone. Chromite.
1915	401,288	1,353									7,500	5,260 tons	72,560	Lime and limestone. Silica.
1916	361,821	1,496	a		a		a				12,000	886 tons	19,613 / 1,717	Silica.
1917	24,758	85	18,982	5,182	a		a				6,200	8,319 tons	1,480 / 167,950 / 104,851	Copper and soapstone. Chromite. Lime and limestone.

Table 1. *Mineral production of El Dorado County, 1880-1953.—Continued.*

Year	Gold value	Silver value	Copper Pounds	Copper Value	Lime Tons	Lime Value	Limestone Tons	Limestone Value	Slate Squares	Slate Value	Misc. stone Value	Misc. stone Amount	Misc. & unapportioned Value	Substance
1918	28,352	722	22,250	5,498			96,673	218,120				2,684 tons	1,506 70	Silica. Other minerals.
1919	30,121	279					41,025	112,423			20,500	11,936 tons	674,856 11,236 6,510	Chromite. Pyrites, silica, soapstone. Chromite.
1920	13,379	155						139,873			1,700 5,500	378 tons 1,600 tons 2,640 tons	13,950 1,169 18,200	Soapstone and talc. Other minerals. Soapstone.
1921	34,109	301					15,296	66,143			2,750	1,498 tons	9,325	Other minerals.
1922	47,340	376					42,200	113,700			4,250	1,652 tons	9,453 18,850	Tale. Slate and soapstone.
1923	30,264	185					95,274	163,987			5,900	2,670 tons	15,729	Soapstone.
1924	28,207	153	[3]		[3]		112,156	322,955	[3]		2,538		8,988 32,691	Tale. Copper and lime.
1925	40,212	238					228,293	297,127	[3]		10,305		4,946	Lime and silica.
1926	91,789	472					59,386	186,702			17,510		5,613	Lime, silica, slate.
1927	82,251	383					96,733	146,506			500		15,792	Copper, gems, silica, soapstone, slate.
1928	122,017	697	1,074	155	[3]		57,012	158,252	[3]		17,455	365 tons	8,455	Soapstone.
1929	57,680	236			[3]		71,033	199,989	[3]		25,665		21,995 83,930	Lead, silica, slate. Copper, lime, silica, slate, soapstone.
1930	78,019	250	[3]		[3]		88,869	205,225	[3]		96,509		113,105	Lead, lime, silica, slate, soapstone.
1931	85,322	283	[3]		[3]		79,798	207,594	[3]		37,494		107,212	Chromite, copper, lead, lime, silica, slate, soapstone.
1932	182,043	438	850	54	[3]	85,938	105,094	207,241	[3]		[3]		97,126	Lead, lime, platinum, silica, slate, soapstone, miscellaneous stone, tungsten ore.
1933	540,989	1,458	2,755	176	[3]		120,026	208,049	[3]		7,551		90,586	Lead, lime, slate, soapstone.
1934	1,380,710	6,035	4,312	345	8,250		112,237	152,422	[3]		7,400		18,405	Lead, silica (quartz), slate, soapstone.
1935	1,803,368	5,943	12,391	1,028	[3]		151,814	298,867	[3]		46,886		232,907	Lead, lime, mineral water, silica (quartz), slate, soapstone.
1936	1,988,735	9,063	21,661	1,963	[3]		159,134	348,055	[3]		77,778		371,356	Chromite, lead, lime, mineral water, platinum, slate, soapstone.
1937	1,719,795	8,238	65,353	7,968	[3]		227,721	448,130	[3]		20,784		102,762	Chromite, lime, mineral water, platinum, slate, soapstone.

Year													Principal commodities
1938	1,484,805	5,717	40,555	3,972	[3]		135,112	304,120	[3]		64,262	343,983	Chromite, lead, lime, mineral water, soapstone, slate.
1939	2,520,105	8,627	10,910	1,135	[3]		146,625	320,212	[3]		16,422	410,654	Lead (224; 4,766 lbs.). Chromite, lime, platinum, mineral water, slate, soapstone.
1940	1,311,585	3,799	1,630	184	[3]		261,713	308,708	[3]		12,947	427,272	Chromite, lead, lime, slate, soapstone.
1941	1,577,630	4,216	957	113	[3]		75,631	152,300	[3]		9,241	580,574	Chromite, lead, lime, slate, soapstone.
1942	636,790	1,624	[3]				147,469	247,522	[3]		15,396	418,918	Chromite, copper, lead, platinum, slate, soapstone.
1943	5,040	303	20,282	2,637	[3]		[3]			[3]		296,469	Chromite, lead, limestone, slate, soapstone, miscellaneous stone.
1944	2,870	574	52,648	7,109						[3]	[3]	288,306	Chromite, limestone, lead, slate, miscellaneous stone.
1945	23,660	791	45,284	6,113						[3]	[3]	271,063	Chromite, lead, limestone, slate, soapstone, miscellaneous stone.
1946	95,305	1,065		[3]	[3]		248,492	578,136			101,170	62,558	Copper, lead, slate, soapstone.
1947	104,760	1,845	112,000	23,520	[3]		[3]	[3]			483,991	1,026,574	Lime, sand and gravel, slate, soapstone, lead and zinc.
1948	90,720	1,259	72,000	15,624	[3]		[3]	[3]			460,086	1,161,110	Lime, sand and gravel, slate and soapstone.
1949	108,990	712			[3]		[3]	[3]			440,674	999,648	Lime, sand and gravel, slate and soapstone.
1950	135,100	713			[3]		[3]	[3]			[3]	1,694,889	Lime, limestone, sand and gravel, stone, slate, soapstone.
1951	132,230	996	3,700	895	[3]		[3]	[3]			[3]	1,896,610	Limestone, lime, slate, stone, soapstone, lead, sand and gravel.
1952	206,675	1,901			[3]		166,106	498,927			[3]	1,270,524	Chromite, lead, lime, sand and gravel, slate, miscellaneous stone, and soapstone.
1953	260,085	2,655	158,000	45,346	[3]					[3]	113,572	1,570,211	Chromite, lead, lime, limestone, slate, soapstone, zinc.
GRAND TOTALS	$21,385,106	$161,706	$834,562	$155,151	51,284	$329,382	3,118,806	$6,634,018	58,611	$181,910	$2,204,530	$15,770,984	

Grand total value, $57,122,787.

(a) Until 1938, a large tonnage of limestone was shipped annually from El Dorado County for use in cement manufacture. Value of this material was included in the state total for cement.

(2) Includes crushed rock, rubble, riprap, sand, gravel.
(3) See under **Unapportioned.**
(4) There was small production of quicksilver in the 1860's, but no record of amounts.
(5) Includes limestone.

CHROME - BEARING   AREAS

1. Flagstaff Hill	3. Latrobe	5. Pilot Hill
2. Clarksville	4. Webber Creek	6. Garden Valley
	7. Georgetown - Volcanoville	

FIGURE 5.   Map of western El Dorado County showing the distribution of serpentine and chrome-bearing areas. (After Cater and others, 1951.)

and Georgetown-Volcanoville areas. During World War II, chromite deposits of El Dorado County were studied in detail by the U. S. Geological Survey (Cater and others, 1951).

Chromite is deposited as magmatic segregations in ultrabasic igneous rocks such as pyroxenite, peridotite, and dunite or in serpentine derived from these rocks. The ore bodies are kidney-shaped masses, pods, elongate lenses, or "leopard" ore (disseminated masses composed of closely spaced spheroids and ellipsoids).

Chaix Mine. Location: sec. 14, T. 8 N., R. 9 E., M. D., 2 miles southeast of Latrobe. Ownership: R. H. Chaix, Box 87, Placerville.

The Chaix chrome mine was worked on a small scale during World War I (Bradley and others, 1918, p. 132) and in 1943 (Cater and others, 1951, p. 124). During the first half of 1953, chrome ore was produced by David Chalmers and Thomas Davidson who sub-leased the property from El Dorado Chrome Company. The mine, which has yielded several hundred tons of chrome ore, has been idle since the summer of 1953.

The deposit consists of disseminated and massive chromite in serpentinized dunite (Cater and others, 1951, p. 124). The chromite bodies, which are discontinuous, commonly pinch and swell, and average $1\frac{1}{2}$ feet in thickness. They occur in a zone that strikes N. 20° E. and dips 70° SE.

During 1953 disseminated ore was mined by bulldozers from an open pit, 300 feet long by 200 feet wide and about 30 feet deep. The Cr_2O_3 content of the ore was about 20 percent (David Chalmers, personal communication, 1953). The ore was trucked 16 miles to El Dorado Chrome Company's custom mill at the Church mine. Mill concentrates were trucked to the Government stockpile at Grants Pass, Oregon.

Darrington (Gurney) Mine. Location: sec. 33, T. 11 N., R. 8 E., M. D., 7 miles southwest of Pilot Hill. Ownership: George Darrington, Folsom.

The Darrington chrome mine was worked originally during World War I, and several hundred tons of ore were produced (Cater and others, 1951, p. 150). During World War II, the property was operated by J. J. Taylor in conjunction with the nearby Dobbas mine. The yield was 495 long tons of ore, most of which was from the Darrington property (Cater and others, 1951, p. 150).

The chromite is in two zones, each in separate masses of dunite. The zones contain disseminated chromite. Some pods and lenses of chromite also are present. The east ore zone is about 300 feet long and 70 feet wide. Typical analyses of ore from the zone are 11.34, 13.76, and 16.65 percent Cr_2O_3 (Cater and others, 1951, p. 151). According to Cater and others (1951, p. 151), the reserves of the milling-grade ore in the east zone may be about 100,000 tons. Some diamond drilling was done in 1943. The west zone is poorly exposed, and only small amounts of work have been done. The iron content of both zones seems to be somewhat high.

The deposits have been developed by open cuts, four adits totaling 900 feet, and shafts and raises totaling 120 feet.

Dobbas Mine. Location: sec. 22, T. 11 N., R. 8 E., M. D., 2 miles due north of Flagstaff Hill. Ownership: D. J. Dobbas, Auburn.

There was a considerable amount of activity on this property during World War I, and a number of open pits and several shafts and adits were developed (Cater and others, 1951, p. 146). During this time, the property was leased or owned by the Placer Chrome Company which sub-leased part of its holdings to the Union Chrome Company. In 1941, the Rustless Mining Corporation prospected the property. Some ore was produced which was concentrated along with ore from several other deposits in the Volo mill near Placerville. In 1942 and again in 1944, some ore was produced from the property (Cater and others, 1951, p. 146).

The deposit consists of a number of scattered ore bodies lying near the eastern margin of a mass of ultrabasic rock that has been altered largely to the talc-chlorite or talc-serpentine rock. The ore bodies strike north to northwest; some of them contain layers of disseminated chromite 1 or more feet in width. A number of assays show the ore to be high in iron (Cater and others, 1951, p. 148). There are five principal ore-bearing areas (Cater and others, 1951, pp. 144, 145). The deposits are developed by open pits and shallow shafts.

El Dorado Chrome Company. During the spring and early summer of 1953, El Dorado Chrome Company operated a custom chrome mill at the site of the Church gold mine in sec. 12, T. 9 N., R. 10 E., M. D., 6 miles south of Placerville. The 20-stamp mill, which was used formerly to treat gold ore, was renovated by the company to beneficiate chrome ore.

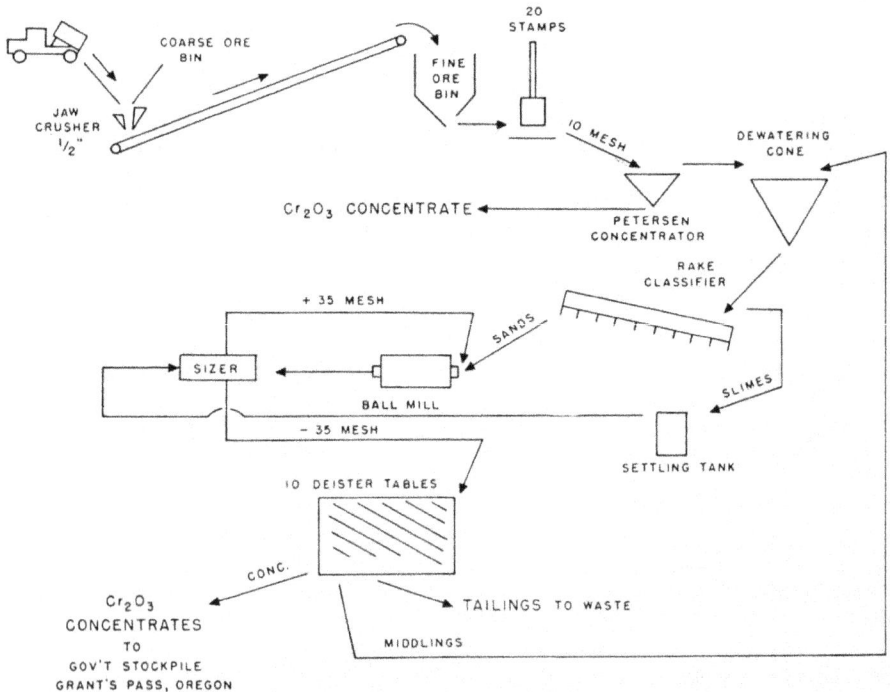

FIGURE 6.   Flowsheet, El Dorado Chrome Company chrome mill, Church mine, March 1953.

This concern leased a number of chromite properties in the Latrobe area, sub-leased them to different operators, and then purchased the ore produced. An approximate base price of 60 cents for each percent of Cr_2O_3 in the ore was paid to the producers. This price varied with each producer, depending on proximity of the mine and mill, and other economic factors. In some cases, the company provided trucks to haul the ore to the mill. The Murphy and Chaix mines supplied most of the ore.

At the mill, the company attempted to maintain an average mill-head grade of 20 percent Cr_2O_3. Mill capacity was 200 tons of ore per day, and mill recovery was about 75 percent. The concentrates were trucked to the Government stockpile at Grants Pass, Oregon.

Joerger Mine. Location: sec. 35, T. 10 N., R. 8 E., M. D., 8 miles west of Shingle Springs. Ownership: Bertha J. Burton, 1115 Yale, Fresno.

Some work was done in this chrome mine during World War I (Cater and others, 1951, p. 126). However, most of the ore from this property was mined in 1942. The exact tonnage is not known as it was included in the combined tonnage from several other mines.

The deposit consists of disseminated chromite found in alternating rich and lean layers. Five zones have been found in serpentine which has intruded metalvolcanic rocks of the Amador group. The zones trend N. 25°-60° E., dip steeply northwest or southeast, and range from 6 to 30 feet in width. A diorite dike has cut off the zones to the east. Estimated ore reserves are 10,000 to 15,000 tons containing 5 to 8 percent Cr_2O_3 (Cater and others, 1951, p. 126). The ore milled in 1942 averaged 8 percent Cr_2O_3. The deposit was worked in an open pit 150 feet long, 15 to 40 feet wide, and 25 feet deep.

Murphy Mine. Location: sec. 14, T. 8 N., R. 9 E., M. D., 2 miles southeast of Latrobe. Ownership: W. C. Egloff, Box 506, Folsom.

The Murphy chrome mine was worked during World War I when about 3000 tons of ore from the property was milled. In 1942 it was worked by the Volo Mining Company when 2000 tons of ore were produced. During the first half of 1953 the mine was operated by Edward Hadsel and Jerry Grant, who sub-leased the property from El Dorado Chrome Company. For a few months in 1953, about 100 tons of ore per day were produced. The mine has been idle since.

The deposit consists of lenses of disseminated chromite in serpentized dunite. The lenses have a general strike of N. 20° E. and dip steeply west. The tenor of the ore mined in 1953 was 14 percent Cr_2O_3 with a 2.7 to 1 chrome-iron ratio (Jerry Grant, personal communication, 1953).

The property is developed by two open pits. The larger pit is 200 feet long and 10 to 30 feet wide with a 25-foot face at the north end. Jackhammers were used for drilling. The ore was trucked 16 miles to El Dorado Chrome Company mill. Eight men were employed at the mine in 1953.

Pillikin (Pilliken) Mine. Location: secs. 21 and 28, T. 11 N., R. 8 E., M. D., in the central part of the Flagstaff Hill chrome area, 6 miles southwest of the town of Pilot Hill and 8 miles due south of

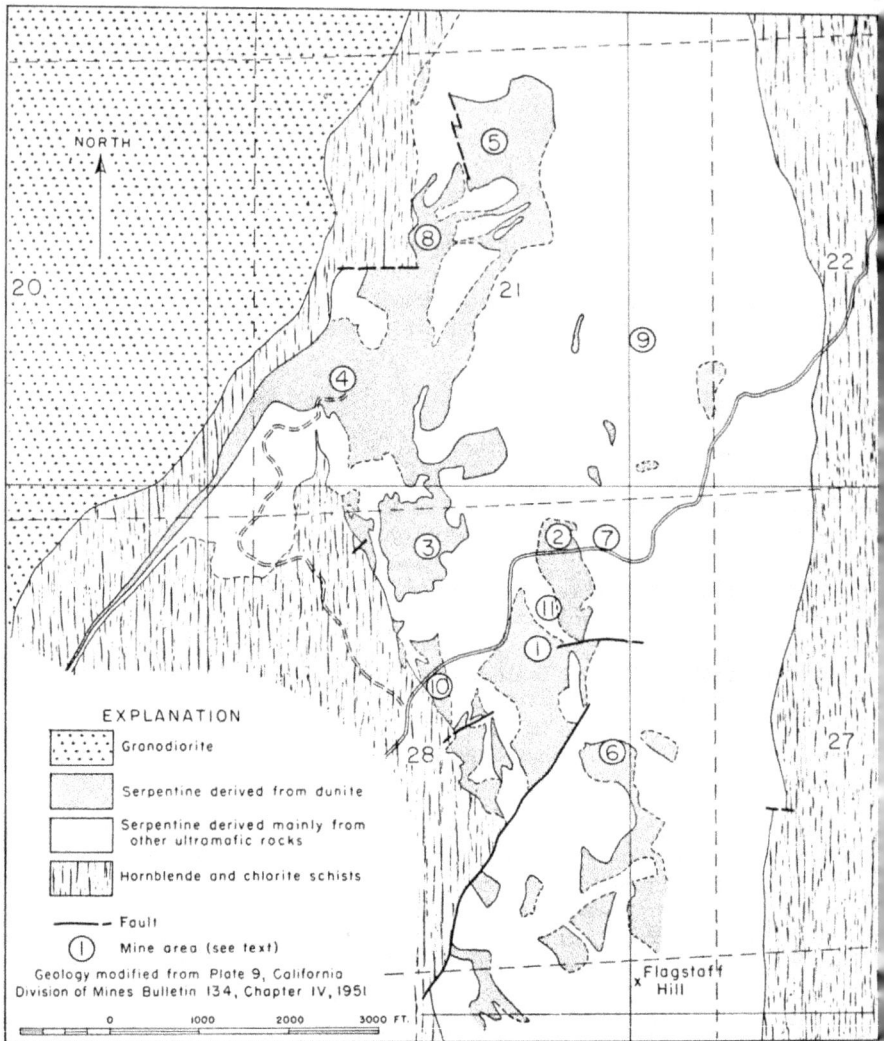

FIGURE 7.   Geologic map showing location of chromite mines, Flagstaff Hill area.

Auburn. Ownership: American Trust Company, 464 California Street, San Francisco.

The Pillikin mine consists of a number of chromite-bearing deposits just north of Flagstaff Hill peak in the extreme western portion of the county. These deposits crop out over an area more than 2 square miles in extent. The mine contains the largest known chromite deposits in the Sierra Nevada and has been the source of more than three-quarters of the total amount of chromite produced in the county (Cater and others, 1951, p. 128).

Chromite was known to exist in this area as early as 1853 (Hanks 1884, p. 136), and a few pockets had been worked prior to 1894 (Lindgren, 1894, p. 4). During World War I, the property was leased by the Noble Electric Steel Company and considerable chromite was pro

FIGURE 8.   Pillikin chrome mine, West Basin Area. Camera facing north.

duced. Most of this output was in lump form that was shipped by rail from Folsom, although a small mill was erected in 1918 and some concentrate made (Cater and others, 1951, p. 131). Also some work was done during World War I by the Placer Chrome Company and the Steele Chrome Company.

The property was idle from 1918 until 1936 when it was acquired by U. S. Chrome Mines, Inc. (Cater and others, 1951, p. 131). This concern erected a 200-ton mill and operated the property until 1939. From 1940 to 1942, the mine was leased and operated by the Rustless Mining Corporation. The property was acquired by the Pillikin Syndicate in 1944. In 1945, J. J. Taylor mined a small amount of ore. Up to 1945, the Pillikin deposits had yielded a total of 27,144 long tons of chromite (Cater and others, 1951, p. 132).

In 1939, the deposit was studied by the U. S. Geological Survey. The result of this work was published in U. S. Geological Survey Bulletin 922-O, in 1940. More detailed studies and a detailed geologic map were made by the U. S. Geological Survey during 1942-43. The results of this later work were published in California Division of Mines Bulletin 134, part III, chapter 4, October 1951.

From 1951 to 1953, the property was leased by the Allied Mining Company, and a new mill was erected. However, little ore was produced and treated, and the mill was dismantled late in 1953. During 1954-55, the mine was leased by the Pillikin Mining Company, Ray Graetz, manager, and some ore was mined from Area 3, West Basin. The ore was treated at the mill at the Pioneer-Lilyama copper mine, 3 miles east of

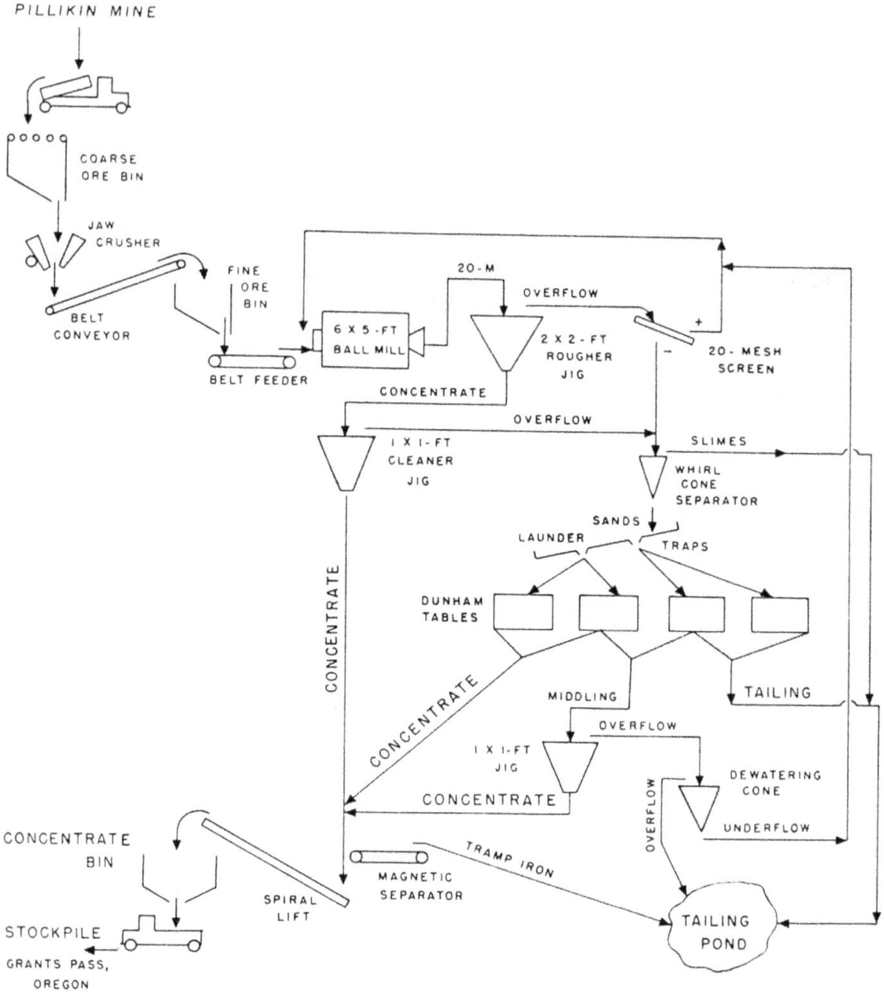

FIGURE 9. Flowsheet of Pillikin Mining Company chrome mill at Pioneer-Lilyama mine, 1955.

the town of Pilot Hill, which had been altered to handle chrome ore. The mine has been idle since April 1955.

The Pillikin chrome deposits are in a mass of ultrabasic rock about three-quarters of a mile wide and 4 miles long. The mass trends north-northwest and dips steeply east. The ultrabasic rocks have been highly serpentinized and otherwise altered. However, four varieties of the original rock have been recognized and mapped: dunite, lherzolite, pyroxenite, and layered olivine and pyroxene rock (Cater and others, 1951, p. 132). Dunite is the most common rock of the mass, while lherzolite is abundant to the east. These rocks are cut by several systems of diorite dikes.

There has been widespread hydrothermal alteration of the ultrabasic rocks which is of economic importance. Four types of alteration are recognized: serpentinization, silicification, formation of talc, and formation of magnesite (Cater and others, 1951, p. 133). Silicification produced rocks ranging from slightly silicified dunite to deep-red jasper.

Chromite is distributed throughout the dunite in concentrations ranging from less than half to more than 30 percent Cr_2O_3. Most of the deposits that have been worked are zones in which the ore consists of alternating chromite-rich and chromite-poor bands. There are a few zones containing evenly distributed disseminated chromite. The lenses and bands of pure chromite range from less than an inch to as much as 4 feet in thickness. The ore zones generally parallel the trend of the ultrabasic rock. The chromite concentrations vary greatly in size ranging from those containing a few pounds to masses containing 300,000 tons (Cater and others, 1951, p. 135). The chromites in the various ore bodies have only small variations in composition (Cater and others, 1951, p. 135). Up to 1945, shipping grade ores averaged 43 percent Cr_2O_3, while concentrates contained 43 percent Cr_2O_3 with a chromium-iron ratio ranging from 1.3 to 2.3.

Because of the irregular shape and erratic distribution of the ore bodies, the potential reserves of the deposit are unknown. Also, the geologic controls of the ore bodies are unknown as yet. It has been estimated that there are at least 450,000 tons of material containing 5 percent or more of Cr_2O_3 at depths permitting open-pit operations (Cater and others, 1951, p. 137).

There are 11 separate chromite-bearing areas in this deposit.

In Area 1 (East Basin), irregular lenses of nearly pure chromite and disseminated chromite are found in light-colored serpentine with talc and small amounts of magnesite. The ore zones trend N. 10° W. and dip 60° E. The area is underlain by dunite with pyroxenite to the east and west. During World War I, the Rustless Mining Corporation produced 37,110 tons of ore averaging 11 percent Cr_2O_3 (Cater and others, 1951, p. 138). In 1952, the open pit was dewatered by the Allied Mining Company and some shipping-grade ore was produced from a lens several feet wide at the base of the pit. The open pit measures about 400 by 400 feet and is about 75 feet deep.

The Area 2 (Bonanza King Mine) portion of the deposit first was worked during World War I by the Noble Electric Steel Company, which shipped 2,500 tons of 46 to 47 percent ore (Cater and others, 1951, p. 138). The mine also was worked by the U. S. Chrome Mines, Inc., which shipped 217 long tons of 43 percent Cr_2O_3 ore, and the Rustless Mining Company, which produced 16,130 tons averaging 9 percent Cr_2O_3 ore. The Allied Mining Company produced some ore from underground workings in this area in 1952. The ore consists of stringers of disseminated and nearly massive chromite. One large lens of ore mined during World War I was 90 feet long and 1 to 12 feet wide. It is developed by two large open pits and some underground workings that are north of the pits.

The Area 3 (West Basin) portion of the deposit was worked extensively during World War I and again during World War II. The Rustless Mining Corporation produced 5,937 tons of ore during 1941-42. In 1952, the Allied Mining Company reopened some of the old underground workings, although there was little ore production.

The last activity at the Pillikin mine—by the Pillikin Mining Company in 1954-55—was in the north portion of this area of the deposit. Two northwest-striking lenses of chromite several feet in width were worked through a 100-foot shaft. Some shipping-grade ore was produced, which averaged 47 percent Cr_2O_3 with a chrome to iron ratio of 2.8 (Ray Graetz, personal communication, 1954). The milling-grade ore, which averaged about 20 percent Cr_2O_3, was milled at the Pioneer-Lilyama mill.

Fine-grained green dunite is the predominant rock type in this area with smaller amounts of pyroxenite and lherzolite. All of the fine-grained green rock may not be dunite, however, as thin-sections made of several specimens taken from this portion of the deposit were found to be fine-grained tactite composed chiefly of epidote. Also garnierite, a nickel-bearing mineral, has been found in some of the laterite soils in this portion of the deposit that have resulted from weathering of the serpentine. As yet, the extent and amount of garnierite present is unknown.

The Area 4 (Chrome Gulch) portion of the deposit was prospected by U. S. Chrome Mines, Inc. in 1942 (Cater and others, 1951, p. 141). A number of large and small chromite bodies occur in partly serpentinized dunite. The chromite is layered; the layers trend north and dip both east and west. There are two major ore bodies that average 5 percent Cr_2O_3. These have been mapped in detail and are shown on plate 11, Bulletin 134, Part III, Chapter 4. The east ore body is estimated to contain at least 250,000 tons of ore and the west ore body 45,000 tons of ore (Cater and others, 1951, p. 142).

During the period 1917-20, *Area 5*, held by the Placer Chrome Company, shipped 7,270 long tons of lump ore and concentrates (Cater and others, 1951, p. 141). Lenses and pods of massive chromite occur in dunite that is interfingered with schist. A north-trending chromite-bearing zone about 1000 feet long is developed by glory holes, shafts, and open pits.

Area 6 was prospected during World War I by the Steele Chrome Company and again around 1940, but little or no ore was produced. Bodies of disseminated ore with a few high-grade streaks occur in a mass of dunite. Exposures indicate at least 10,000 tons of ore containing 5 percent or more of Cr_2O_3 (Cater and others, 1951, p. 143). The deposit is developed by open cuts and short adits.

Small amounts of ore were produced from *Area 7* during World War II. There is an estimated total of 25,000 tons of disseminated ore in a talcose dunite body 200 feet wide and at least 1,500 feet long (Cater and others, 1951, p. 143). It is developed by open pits and trenches.

The deposit in *Area 8* consists of dunite containing high-grade layers of chromite as much as 24 inches thick. The enclosing dunite contains some disseminated chromite. It is developed by several open pits and short adits.

During World War I, *Area 9 (Nillson or Donnelly mine)* shipped 224 long tons of ore (Cater and others, 1951, p. 143). The area was prospected again around World War II. Irregular lenses of chromite up to 18 inches in thickness occur in dunite that is almost completely altered to talc. Disseminated ore is also present.

In *Area 10* disseminated ore containing a few high-grade streaks occurs in a northwest-trending mass of dunite. Although some good ore is present, the average grade apparently is too low and the ore bodies too small for profitable mining (Cater and others, 1951, p. 146).

In *Area 11* several lenses of high-grade chromite as much as 3 feet thick were mined from shallow pits, inclined shafts, and adits. There is a high degree of shearing and much silification.

Pillikin Mining Company Mill. Location: sec. 3, T. 11 N., R. 9 E., M. D., at the Pioneer-Lilyama copper mine. Ownership: H. H. Mitchel, 9490 Brighton Way, Beverly Hills.

From November 1954 to April 1955, the Pioneer-Lilyama was leased by the Pillikin Mining Company. Chrome ore from the Pillikin mine was concentrated and the concentrates trucked to the Government stockpile at Grants Pass, Oregon. Ore was trucked to the mill from the Pillikin mine, 8 miles to the southwest, and treated by crushing, fine grinding, jigging, and tabling.

Some of the milling equipment that was used previously in the treatment of copper ore from the Pioneer-Lilyama mine was utilized. The ball-mill product was 20 mesh. Water from milling was obtained from the adit of the copper mine. Fifty tons of ore per day were treated. Three tons of ore yielded an average of about 1600 pounds of concentrates (Gratz, R., personal communication, 1955). Mill heads averaged 27 percent Cr_2O_3. Two shifts of four men each worked at the mill.

Walker Mine. Location: sec. 35, T. 10 N., R. 8 E., M. D., 8 miles west of Shingle Springs. Ownership: Faustino Silva, Route 8, Box 951, Sacramento.

Some chrome ore was produced from this property in 1917 and again in 1918. In 1942 the property was worked by the Volo Mining Company of Placerville (Cater and others, 1951, pp. 126, 128).

The deposits consist of disseminated chromite in alternating rich and lean layers in serpentine which was intruded into metavolcanic rocks of the Amador group. The ore zone strikes N. 35° W., dips steeply northeast, and is 3 to 4 feet wide. Estimated ore reserves to the depth of the shaft are 1500 tons containing 10 to 12 percent Cr_2O_3. The property is developed by a 60-foot shaft with 30 feet of drifts on the 60-foot level and numerous open cuts.

Copper

Recorded output of copper in El Dorado County since 1880 is 834,562 pounds, valued at $155,151. However, copper was produced in the county prior to that date. In 1859, copper ore was being mined at the Cosumnes mine, and it was known to occur in several other localities in the county (Aubury, 1908, p. 31). During the Civil War days of the early 1860's, substantial quantities of copper ore were produced in the Sierran foothills and shipped to smelters in Boston, Baltimore, and Wales. During World War I, copper was produced at several properties. Since 1927, some copper ore has been produced in the county nearly every year. At the present time there is some production from the El Dorado, Noonday, Pioneer-Lilyama, and Cosumnes mines. Also, copper ore from the Copper Hill mine in Amador County is concentrated in the Volo mill. Small amounts are recovered as by-products of gold mining.

Along the western Sierran foothills and west of the Mother Lode is a belt of copper and zinc mineralization. This belt contains lenticular sulfide bodies that were formed by cavity filling and replacement along zones of shearing, faulting, and crushing in metamorphic rocks of Paleozoic and Jurassic age. The ore bodies are composed chiefly of pyrite with varying amounts of pyrrhotite and chalcopyrite. Sphalerite, bornite, chalcocite, galena, tetrahedrite, gold, and silver are present in smaller amounts. The ore deposition is controlled by the structure, and the copper mineralization is genetically related to the emplacement of the Sierra Nevada granitic batholith.

Copper in El Dorado County also occurs in contact metamorphic rocks where granitic rocks have intruded calcareous metamorphic rocks. Chalcopyrite with appreciable amounts of bornite are the principal ore minerals in these deposits. Copper ore is found in the Mother Lode belt at the El Dorado copper mine at Garden Valley. Copper also occurs as a minor constituent of lode gold ores.

Big Buzzard (Hercules, Darrington) Mine. Location: sec. 29, T. 11 N., R. 8 E., M. D., 3 miles southwest of Rattlesnake Bridge and half a mile east of the American River. Ownership: George Darrington et al., Folsom.

The Big Buzzard copper and zinc mine originally was worked as a gold mine, and the ore was treated in a five-stamp mill (Logan, 1923, p. 142). During the period 1900-10, the mine was leased by several concerns; however, there was little production. In 1926 there was some additional underground development (Logan, 1926, p. 406). During 1943 and 1944, several lessees shipped some copper-zinc ore from the dump. In 1948, the Morning Star Mining Corporation leased the mine and made a test run of some dump material at the Pioneer-Lilyama mill.

The deposit occurs in amphibolite and mica schist near granodiorite which lies to the west. The vein strikes northwest and dips northeast. It is as much as 10 feet wide. The ore is complex and contains appreciable quantities of sphalerite with pyrite, chalcopyrite, and galena. Some of the ore contained as much as $14 per ton in gold (Logan, 1926, p. 406). The ore shoot pitches south. There is a 300-foot inclined shaft sunk on the vein with levels at 70, 160, 260, and 300 feet. Most of the stoping has been on the 70 and 160-foot levels.

Cosumnes Mine. Location: secs. 24, 25, T. 9 N., R. 12 E., M. D., 4 miles northwest of Fairplay, by the Cosumnes River. Ownership: Clifford Smith, Somerset, leased by Lee Miller, Fairplay.

The Cosumnes copper mine was worked originally in 1859 (Aubury, 1908, p. 31). It was active again around 1896 and during World War I. During World War II, a small amount of ore was produced from open cuts on the hill above the adit portal and treated at the Volo mill near Placerville (Lee Miller, personal communication, 1955). The present operator, who leased the property early in 1955, has reopened some of the old workings and produced several tons of ore that have been stockpiled.

Copper minerals occur in a contact metamorphic zone that lies between limestone on the east and granodiorite on the west. The principal gangue is tactite composed of garnet and epidote with some calcite and quartz. The ore minerals are bornite and chalcopyrite. Pyrite, molybdenite, magnetite, and powellite are present in smaller amounts. Minor amounts of gold and silver are present. The tactite bodies trend northeast and are as much as 30 feet wide. Other rocks in the immediate area are amphibolite and mica schist. Some of the tactite is slightly radioactive (Lee Miller, personal communication, 1955).

The mine is worked through a 150-foot crosscut adit driven westward with drifts running northeast and southwest along the mineralized zone. A 30-foot raise is being driven just south of the adit-drift intersection. Also, there is a lower crosscut adit about 40 feet below the main adit. These two levels are connected by a winze. There are a number of caved workings the extent of which are unknown. There are numerous open cuts on the hill above the adit portals. Three men including Mr. Miller work at the mine.

FIGURE 10.   El Dorado copper mine.

El Dorado (Roosevelt) Mine.   Location: sec. 34, T. 12 N., R. 10 E., M. D., 1 mile southeast of Garden Valley. Ownership: Calivada Development Company, Box 4, Garden Valley.

Since 1953, there has been intermittent development work and some ore production from El Dorado copper mine. The shaft was deepened to 100 feet and about 50 feet of drifts were driven on the new 100-foot level. New surface equipment, including a headframe, buildings, and a compressor, were installed. This work is being done by the Calivada Development Company, H. T. Hall, president.

This mine originally was worked for gold in the 1860's when a north-trending adit was driven. During World War II, some copper ore was mined. In 1944-45, the U. S. Bureau of Mines made an examination of the property, the results of which were published in the U. S. Bureau of Mines Report of Investigations 3896, June 1946 (Bedford, 1946). Exploration was done by means of diamond drilling and sampling of exposures.

El Dorado copper deposit is in the Mother Lode gold belt. Country rock is black Mariposa slate. Small amounts of fine-grained green tuffaceous rock are interbedded with the slate. The ore, which is indicated on the surface by narrow zones of gossan, is composed of massive chalcopyrite, pyrrhotite, and pyrite. The massive sulfides occur in bands up to several feet in thickness and evidently have replaced the green tuffaceous rock. There is also disseminated chalcopyrite in the wall rock of the deposit. On the 100-foot level there are five parallel bands of ore in a zone about 5 feet wide. Strike is to the northwest; dip is to the northeast or vertical.

Exploration work done by the U. S. Bureau of Mines consisted of eleven diamond drill holes aggregating a total of 1,613 feet. Eight of the holes drilled indicated the deposit to be essentially a series of narrow intermittent lenses of ore along structures about 600 feet in length (Bedford, 1946, p. 1) and extending several hundred feet in depth. Apparently the greatest amount of ore is in the vicinity of the shaft. Copper content of the ore ranges from 5 to a little more than 10 percent. As much as 1.5 percent zinc, about 1 ounce per ton in silver, and traces of nickel and gold also are present in the ore.

The mine is developed by a two-compartment 100-foot shaft inclined to the east. A north-trending 173-foot adit, driven as a crosscut for 46 feet and as a drift for 127 feet, connects with the shaft on the 50-foot level 128 feet in from the adit portal. On the 100-foot level there are drifts extending 35 feet to the north and 10 feet to the south.

Noonday Mine. Location: sec. 18, T. 9 N., R. 11 E., M. D., 4 miles southeast of El Dorado and half a mile east of the Mother Lode. Ownership: George Fausel, Placerville, leased by the Noonday Copper Mining Co., R. F. Fitzgerald, president.

The Noonday copper mine was active during the period 1900-05 when it was developed by shallow workings (Eric, 1948, p. 231). A shaft later was extended to a depth of 200 feet, and several hundred feet of drifts were driven. Ore containing 5 to 9 percent copper and some gold and silver was mined (Tucker, 1919, p. 278).

Early in 1953, the mine was reopened by the Noonday Copper Mining Company. The shaft was rehabilitated and about 100 feet of new drifts driven. Several thousand tons of ore was mined and was milled at the Volo mill west of Placerville. It was shut down in 1954. Early in 1956 the mine was reopened, and at present small amounts of ore are being mined from the 200-foot level.

This deposit occurs in a shear zone in schist of the Calaveras group, about half a mile east of the Mother Lode gold belt. The vein strikes north to northeast and dips steeply east. It is as much as 7 feet wide. The ore consists of chalcopyrite associated with pyrite and smaller amounts of bornite. Ore mined during the last operation contained as much as 4 percent copper, although the tenor was considerably less

FIGURE 11. Volo mill at Shaw gold mine.

(Joseph Pickering, personal communication, 1955). Minor amounts of gold and silver were recovered. The mine is developed by a two-compartment 230-foot vertical shaft with levels at 100 and 200 feet. There are approximately 350 feet of drifts. Five men work at the mine.

Pioneer-Lilyama (Little Emma, Volo) Mine. Location: sec. 3, T. 11 N., R. 9 E., M. D., 3 miles east of Pilot Hill. Ownership: H. H. Mitchel, 9490 Brighton Way, Beverly Hills; leased by Wilcox-Lilyama Mining Company, F. R. Wicks, manager, Frank James, mine foreman, and D. W. Reynolds, mill foreman.

This property is a consolidation of the Pioneer and Lilyama copper mines, which originally were worked in the 1860's (D. W. Reynolds, personal communication, 1954). However, the main copper deposit was not discovered until 1889-90 when the adits were driven (Aubury, 1908, p. 212). The mine was idle until 1943 when it was reopened by the Volo Mining Company of Placerville (Cox and others, 1948, p. 44). A considerable amount of copper and smaller amounts of gold and silver were produced until 1948, when the mine was shut down. The ore was concentrated by flotation in a mill on the property. During 1943-44, the U. S. Geological Survey examined the ore deposit and made geologic maps of the area. The results of this work were published in California Division of Mines Bulletin 144.

During the fall of 1954 and spring of 1955 the mill was leased to the Pillikin Mining Company and used to treat chrome ore from the Pillikin mine. Since July 1955 the present operator has been rehabilitating the mine and altering the mill. Some ore has been produced and is stockpiled, awaiting completion of the work on the mill.

This deposit is near the west margin of a granodiorite stock that has intruded metasedimentary rocks of the Calaveras group. The deposit trends north and is in contact-metamorphic rocks, which include tactite and hornfels and marble. Hornfels is the most prevalent metamorphic rock type near the mine, while granodiorite is the principal intrusive rock. The tactite contains garnet, epidote, magnetite, quartz, calcite, and smaller amounts of idocrase, pyrite, chalcopyrite, bornite, ilmenite, feldspar, and specular hematite. Minor amounts of scheelite have been found (Cox and others, 1948, p. 45). The composition of the tactite varies considerably. The tactite bodies range in size from small pods to those 100 or more feet in width.

The ore bodies are those parts of the tactite that contain commercial amounts of chalcopyrite and bornite. The sulfides occur as irregular masses or veinlets and commonly are associated with abundant euhedral magnetite crystals. Small amounts of gold and silver are present in the sulfides. The mine is developed by four northeast-trending crosscut adits. An upper adit encounters ore 120 feet in from the portal, and a north-trending 150-foot drift has been driven along the ore zone. On the hill above are open cuts, trenches, and two glory holes which are connected with this level by raises. The adit has been rehabilitated, but no work is being done on this level.

On the lower or main level, there are two nearly parallel northeast-trending crosscut adits about 30 feet apart. The southern adit, which is the main haulageway, encounters the ore zone 250 feet from the portal. The ore zone was explored by a 100-foot drift to the southeast and a 120-foot drift northwest. From the end of the northwest adit, a crosscut was driven 180 feet northwest to a second ore zone that was explored by short drifts. At the intersection of the adit and drifts, a 63-foot vertical winze connects with a lower level. On this lower level, drifts extend 35 feet northwest and 75 feet southeast. Also, there are several inaccessible shafts on the property. Present work is confined to the winze and the drifts southeast of the adit. These are being rehabilitated, and any ore produced is stockpiled at the mill.

The mill is equipped with a Wheeling jaw crusher, conveyor belt, 6- by 6-foot Allis Chalmers ball mill, Wemco screw classifier, drum-type magnetic separator, 10 Denver sub-flotation cells, and two Dunham Economy tables. The capacity of the mill is being increased from 150 to 200 tons per day (D. W. Reynolds, personal communication, 1956). In 1948, the concentrate averaged 28 percent copper, 28 to 34 dollars per ton in gold, and 14 dollars per ton in silver (D. W. Reynolds, personal communication, 1956).

Rip and Tear (Dodson) Mine. Location: sec. 3, T. 8 N., R. 9 E., M. D., 2 miles north of Latrobe. Ownership: F. H. Dodson, 2404 26th Street, Sacramento.

The Rip and Tear or Dodson copper mine was worked originally during California's "copper boom" of the 1860's. In 1918, the shaft was reopened and two carloads of ore were shipped to a smelter (Logan, 1926, p. 408). In 1943, the property was leased by W. J. Varozza of Latrobe who cleaned out the underground workings and shipped a small amount of ore (F. H. Dodson, personal communication, 1954). It has been idle since.

The deposit consists of bands and stringers of massive pyrite containing appreciable amounts of chalcopyrite and pyrrhotite. These occur in a northwest-striking shear zone in fine-grained metavolcanics. The ore body averages 5 feet in width and contains as much as 10 percent copper (F. H. Dodson, personal communication, 1954). Appreciable quantities of gold and silver also are present. The mine is developed by a 100-foot main shaft and drifts. There is a 40-foot shaft on the same property about a mile to the north of the main shaft.

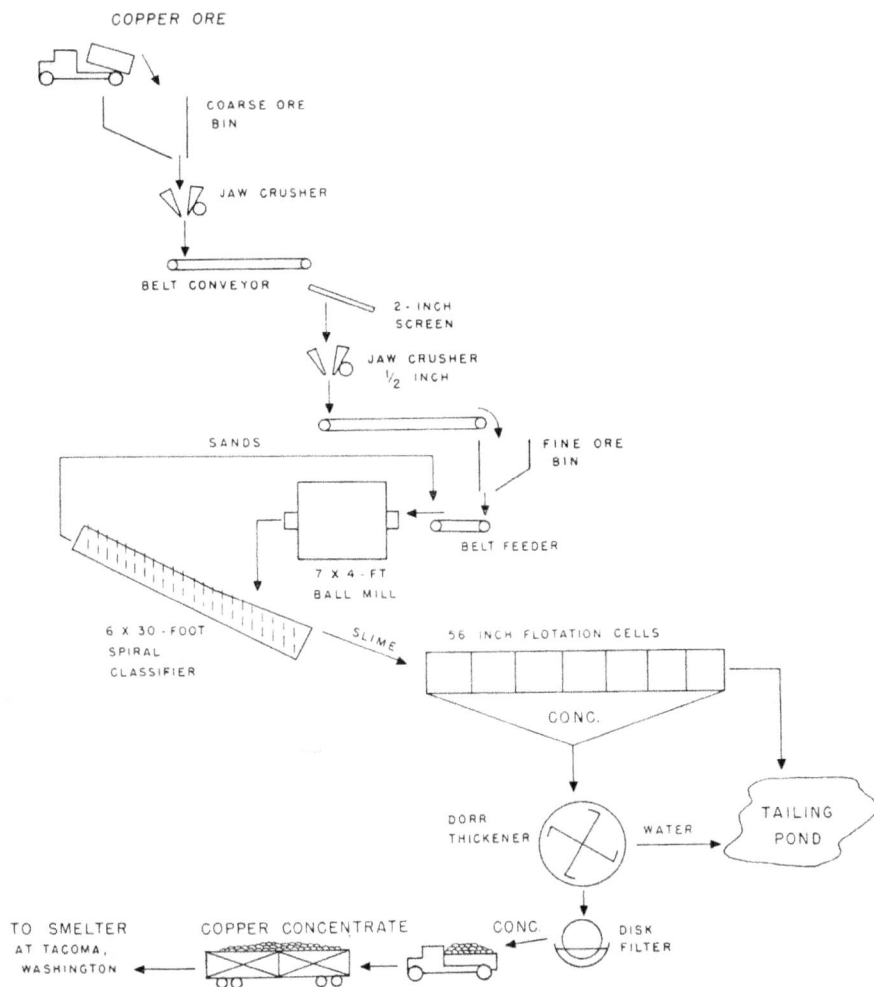

FIGURE 12.   Flowsheet (copper circuit) at Volo mill.

Volo Mill. Location: sec. 21, T. 10 N., R. 10 E., M. D., at Shaw mine, 5 miles southwest of Placerville. Ownership: Volo Mining Company, Placerville, F. V. Phillips, president, J. Pickering, mill foreman.

During 1941-42 and again from 1946-53, gold ore from the nearby Shaw mine was treated in this mill by gravity concentration, flotation, and cyanidation. During World War II, copper ore from several properties in the county was treated in the mill by flotation. Also some chrome ore was milled during World War II. In 1953, copper ore from the Noonday mine was milled. The Shaw and Noonday mines, however, were shut down and the mill was idle until the fall of 1954, when the Copper Hill copper mine in Amador County was reopened. New equipment was installed and the flowsheet simplified. At the present time, about 50 tons of copper ore per day from the Copper Hill and Noonday mines are treated by flotation. Concentrates are shipped by railroad to the American Smelting and Refining Company's Tacoma smelter. Millheads contain 4 to 7 percent copper and 1 to 5 percent zinc (J. Pickering, personal communication, 1955). The zinc is not recovered, although there is a zinc flotation circuit at the mill. Smelter receipts show the concentrates contain 15 to 20 percent copper, 8 to 12 ounces of silver per ton, and 0.1 to 0.2 ounces of gold per ton. Four men work at the mill.

Gold

Introduction. Although the total gold output of El Dorado County is not as great as that of several other counties in California, there is probably no other area of similar size in the state where gold occurs in such a wide variety of deposits. Gold not only occurs in veins in the Mother Lode and the East and West belts, but also in contact metamorphic and replacement deposits. It is found in stream placers ranging from Tertiary to Recent in age, and in residual placers or "seam diggings."

History. California's gold rush began soon after James Marshall's discovery at Coloma in 1848. The rich and virgin surface placers yielded large amounts of gold during the first few years of the gold rush. After these were exhausted, hydraulic mines were a major source of gold in the county. Also "seam diggings" yielded substantial amounts. In 1884, hydraulic mining was severely curtailed by the Sawyer Decision, which prohibited the dumping of debris into the Sacramento and San Joaquin Rivers and their tributaries.

After the Sawyer Decision was passed, underground lode mines became the chief source of gold in the county. During the latter part of the nineteenth century, improved underground mining practices as well as new concentrating methods, such as chlorination and, later, cyanidation, were introduced which made many lode deposits workable economically. During the flush years of World War I and the 1920's, the gold output of El Dorado County was small.

During the 1930's, low costs, coupled with an increase in the gold price to $35 per ounce in 1934, caused gold production figures to rise abruptly. During these times, a large number of creeks were worked by dragline dredges or "doodlebugs." The peak of this production came in 1939 when more than 2½ million dollars worth of gold was produced.

World War II with its high wages and scarcity of materials caused a great drop in gold mining in the county. In November 1942 War Production Board Order L-208 went into effect which further restricted gold mining, and nearly all of the mines closed. Only a few thousand dollars were produced in 1943-44, much of it a by-product of copper-zinc mining.

Order L-208 was lifted in July 1945, but only a few mines reopened. Since World War II, gold mining has continued to follow a diminishing trend. The discovery of the Hazel Creek mine, which now is the chief source of gold in the county, has been the only recent significant development in gold mining in El Dorado County.

Lode Deposits

Lode-gold deposits in El Dorado County are found in the Mother Lode system, the East Belt, and the West Belt. The Mother Lode belt has been the source of the largest amount of lode gold produced in the county. There also has been substantial gold production in both the East and West Belts.

The Mother Lode belt is a system or zone of steeply dipping gold-bearing quartz veins that traverses the central portion of the county. The belt trends north through Nashville, northeast through Placerville, and northwest to Garden Valley. At Garden Valley, the Mother Lode belt splits with one branch extending northwest through Greenwood to the Middle Fork of the American River, while the other extends north through Georgetown to the Georgia Slide area. This area is considered to be the north end of the Mother Lode belt (Knopf, 1929, p. 48). A number of the gold mines in the northern portion of the Mother Lode belt in El Dorado County are classed as "seam diggings" (see section on seam deposits).

The Mother Lode veins are enclosed in Mariposa slate with associated greenstone. The vein system or zone ranges from a few hundred feet to a mile or more in width. Within the zone are numerous discontinuous or linked veins which may be parallel, convergent at small angles, or slightly en echelon. They cut the enclosing rocks at acute angles both in strike and dip (Knopf, 1929, p. 24). Few of the veins can be traced for more than a few thousand feet along the strike. The veins commonly pinch and swell.

Mother Lode veins were formed in fissures that developed in a zone of reverse faulting. Repeated movements along the fissures facilitated the passage of mineral-bearing solutions that originated from deep-seated granitic magmas of the Sierra Nevada. The veins are composed of quartz in which disseminated gold and associated sulfides, chiefly pyrite, occur in ore shoots of varying extent. Other minerals present include ankerite, calcite, chlorite, sericite, and mariposite. The grade of Mother Lode ores is low to moderate with an average of roughly $10 per ton in gold. However, the ore shoots often are extensive and persist at depth, in some cases to several thousand feet. Bodies of auriferous greenstone and altered schist occur adjacent to Mother Lode veins as wall-rock replacements in several areas in the county.

Lying east of the Mother Lode is the area known to miners as the East Belt. It consists of many individual gold-bearing veins in metamorphic rocks of the Calaveras group or in granitic rocks. Although many of these deposits trend north, a few trend west. Most dip steeply. The ore shoots are smaller and the veins narrower than those of the Mother Lode, but commonly they are richer. The gold usually is associated with appreciable amounts of pyrite, chalcopyrite, pyrrhotite, galena, sphalerite, and arsenopyrite. The most productive areas of the

FIGURE 13. Longitudinal map of the underground workings, Alhambra gold mine, showing principal work done by the Alhambra-Shumway Mining Company.

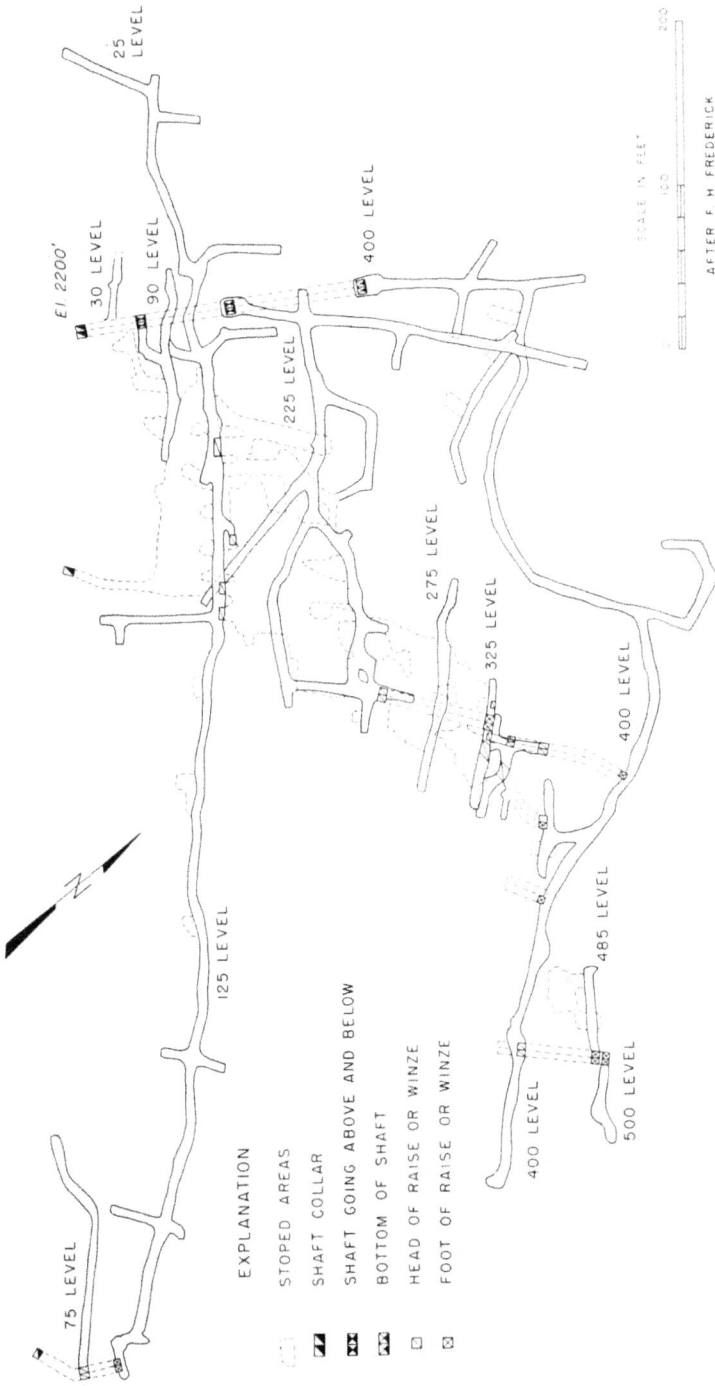

FIGURE 14. Composite map of underground workings of the Alhambra gold mine.

East Belt in El Dorado County have been the Sly Park, Grizzly Flat, and Spanish Flat areas. The Hazel Creek mine north of Grizzly Flat is the chief source of gold in the county at the present.

West of the Mother Lode in the West Belt, the gold is in quartz veins and in bodies of greenstone. The veins occupy fissures in amphibolite and chlorite schist and granodiorite. The gold is associated with pyrite and small amounts of galena. The auriferous greenstone bodies are replacement deposits in shear zones. Much of the gold in these deposits is closely associated with pyrite and in some cases with arsenopyrite. The largest production in the West Belt has been in the Deer Valley, Shingle Springs, and Frenchtown areas.

Alhambra Mine. Location: secs. 6 and 7, T. 11 N., R. 11 E., M. D., 1 mile east of Spanish Flat and 2 miles northeast of Kelsey. Ownership: Wilbur E. Timm, c/o County Courthouse, Placerville; leased by Alhambra-Shumway Mines, Inc., 681 Market Street, San Francisco, and sub-leased to Alhambra Gold Mine Corporation, 1930 Outpost Drive, Hollywood.

The Alhambra gold mine was worked originally in 1883 when a 29-foot shaft developed an ore shoot that yielded $27,600 (Logan, 1938, p. 216). It was active again in 1886, and in 1890 there was a five-stamp mill on the property (DeGroot, 1890, p. 178). The mine was idle until 1934 when it was reopened by Jensen and Schneider. These operators discovered two high-grade pockets at a depth of about 90 feet, each of which yielded more than $10,000 (Logan, 1938, p. 216).

Soon after, the Alhambra-Shumway Mining Company was formed, which leased the property. This concern since has done most of the development work and is responsible for the largest portion of the gold produced from the mine. In 1939, one of the richest pockets of high-grade ore found in California in recent years was discovered in the Alhambra mine. This pocket, in the vein between the 225- and 275-foot levels and about 250 feet southeast of the shaft, yielded about $550,000 (F. H. Frederick, personal communication, 1955). This discovery attracted much attention at the time and was featured in several newspaper articles. The mine was active until late in 1942 when it was shut down.

In 1945, the mine was leased by W. W. Williams who did some prospecting work. However, there was no reported gold production. In 1947, the mine was sub-leased by the Alhambra Gold Mine Corporation, O. H. Griggs, president. The mine was de-watered, equipment rehabilitated, and new development work done. According to O. H. Griggs, additional underground development work was done, most of it to block out ore between the 125- and 360-foot levels (this would be in addition to the work shown on the accompanying mine maps). In 1949, a high-grade pocket was discovered on the 225-foot level. A portion of this ore weighing 197 pounds yielded 324 ounces of gold (O. H. Griggs, personal communication, 1955). Soon after this discovery, the concern became involved in a lawsuit, and the mine was shut down. It is now idle, pending the outcome of the suit. The total output of the mine is approximately $1,250,000 (F. H. Frederick, personal communication, 1955). At least 50 percent of this total was from high-grade ore.

This deposit is about 2 miles east of the Mother Lode and lies near the east margin of the belt of Mariposa slate. Country rock consists of interbedded amphibolite schist, graphitic schist, and slate. The vein occurs in a fault zone and consists of strings of nodule- or pod-like bodies of quartz in chlorite-sericite schist and slaty schist. The vein has an average strike of N. 50° W. and dips northeast. There are several smaller parallel veins in the hanging wall. The main hanging wall is composed of hard amphibolite schist, while the footwall is graphite-sericite schist and slate. The main fault-zone trends northwest and dips northeast. East of the main shaft the main fault splits to the northwest; the two branches continue to strike in a general northwest direction, but 10 to 20 degrees apart. The junction has been favorable for the localization of the ore shoot and also is the channel along which the gold-bearing solutions are believed to have come (F. H. Frederick, personal communication, 1955).

Much of the gold was free and was associated with pyrite and appreciable amounts of arsenopyrite. The ore shoot was lensoid in character and plunged to the southeast. The high-grade pockets also were lensoid in character and plunged to the southeast. The famous pocket found in 1939 was nearly 5 feet wide (see California Division of Mines Bulletin 141, plate 2, view F). Associated with the quartz veins are a number of lensoid bodies of calcite and several schistose dikes.

The mine is developed by a 440-foot inclined shaft and an older 75-foot shaft about 600 feet to the southeast. There are approximately 3000 feet of drifts and crosscuts. The principal levels are the 125-, 225-, and 400-foot levels. About 450 feet southeast of the shaft on the 400-foot level, a 100-foot winze was sunk, and a small ore shoot below the main ore shoot was stoped from the 485-foot sub-level. The surface plant includes a 50-ton mill equipped with jigs, tables, and flotation cells; a steel headframe; and several other buildings.

Alpine Mine. Location: secs. 15 and 16, T. 12 N., R. 10 E., M. D., 2 miles southeast of Georgetown. Ownership: Elizabeth F. Burks, 4240 12th Street, Sacramento.

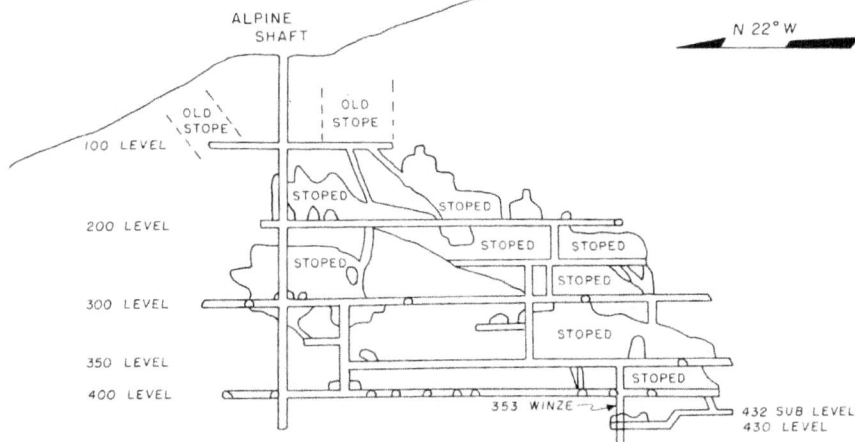

FIGURE 15.   Longitudinal projection of underground workings of the Alpine gold mine.

The Alpine gold mine was worked originally in the late 1860's (Logan, 1934, p. 15). In 1888, a 10-stamp mill was in operation on the property. The mine was active around 1902 and again about 1912 (Logan, 1934, p. 15). From 1933 to 1938, the property was operated by the Beebe Gold Mining Company, and the ore was treated at the Beebe mill. A total of $434,665 was produced from 64,349 tons of ore in this last operation (Seventh annual report of the Beebe Gold Mining Company, 1939).

A gold-bearing vein of white sugary quartz strikes northwest and dips 60° to 65° NE. A number of south-dipping faults cross the vein and flat-dipping ore shoots, about 100 feet long, were found at these intersections (Logan, 1934, p. 15). Most of the gold was free and finely disseminated. Only minor amounts of sulfides are present in the ore. Country rock is amphibolite.

The mine is developed by a 400-foot shaft with levels at 100, 200, 300, 350, and 400 feet. There also is a 430-foot sublevel. The ore was trucked to the Beebe mill at Georgetown.

Argonaut Fraction Mine. Location: sec. 17, T. 12 N., R. 10 E., M. D., by Georgetown Creek, a quarter of a mile northeast of the Argonaut mine and 1½ miles east of Greenwood. Ownership: Pete Lopez, 4833 9th Avenue, Sacramento.

This gold property has been operated intermittently since 1933 by Mr. Lopez. Small amounts of ore are produced and stockpiled at the mine.

Two parallel northwest-striking veins about 150 feet apart have been developed. The ore consists of mineralized green schist with considerable amounts of pyrite. The veins range from 2 to 4 feet in width. One shipment of ore during 1937 averaged more than $15 per ton (Pete Lopez, personal communication, 1955).

A 100-foot drift adit has been driven southeast along the east vein and a 60-foot adit along the west vein. Several hundred tons of ore have been stockpiled. Two men work intermittently at the mine.

Barnes-Eureka (Greenstone) Mine. Location: sec. 33, T. 10 N., R. 10 E., M. D., 2 miles northeast of Shingle Springs. Ownership: B. F. Baskin, Route 2, Box 38, Placerville.

Originally worked prior to 1894 (Crawford, 1894, p. 102), the Barnes-Eureka mine was the source of small amounts of gold in 1912. Some work was done on the property in 1936 (Logan, 1938, p. 217). The property was worked on a small scale again during 1947–49 by B. F. Baskin of Placerville.

Gold associated with arsenopyrite and tellurides is in a 2-foot quartz vein on the contact between serpentine on the east and fine-grained metavolcanic rocks on the west. The vein strikes north and dips 45° E. The property is developed by a 350-foot inclined shaft with levels at 100 and 200 feet and a second 250-foot shaft to the south.

Beebe (East Lode, Brooklyn, Iowa, Woodside-Eureka) Mine. Location: secs. 2, 3, and 11, T. 12 N., R. 10 E., M. D., on the north side of Georgetown. Ownership: Woodside-Eureka Mining Company, Ltd., 1605 Tribune Tower, Oakland 12.

The Beebe mine, which is a consolidation of a number of claims, was one of the larger sources of gold in the county. The Eureka claim

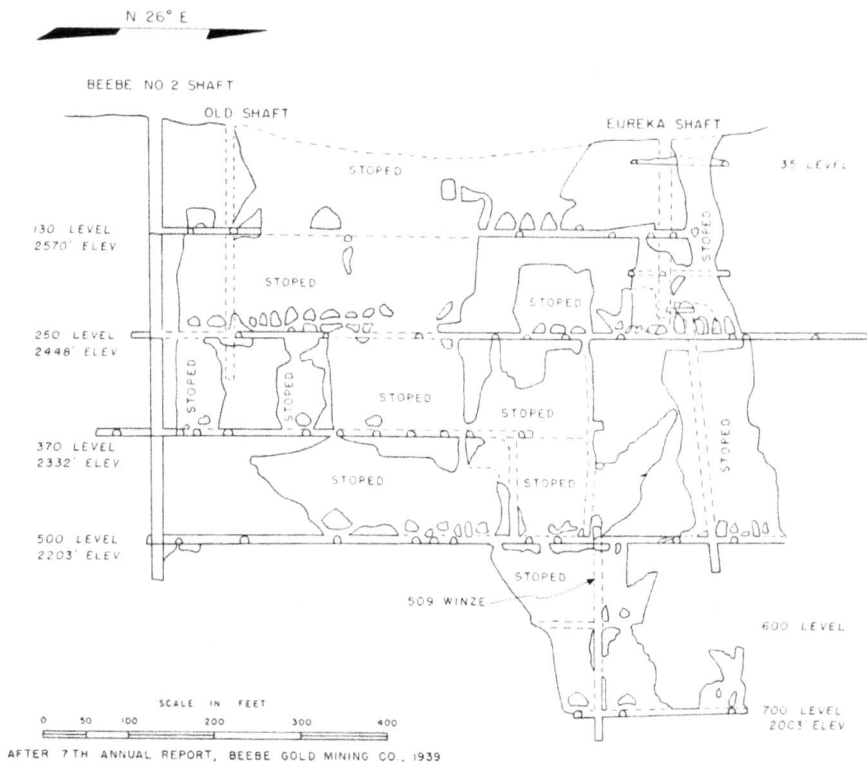

FIGURE 16.   Longitudinal projection of the underground workings of the
Beebe gold mine, showing stoped areas.

was first worked in the early days of the gold rush and later at intervals until 1908 (Logan, 1934, p. 17). The Beebe claim was first prospected in 1917 (Logan, 1934, p. 17). From 1932 to 1939, the property was operated by the Beebe Gold Mining Company. A total of $1,200,465 was produced from 306,241 tons of ore during this operation (Seventh annual report of the Beebe Gold Mining Company for 1939). After 1939, there was only a small amount of gold produced, and it was from cleanup operations. Ore from the Alpine mine also was treated at the mill (see Alpine gold mine).

The Beebe-Eureka vein is a silicified and mineralized zone in amphibolite schist that strikes northeast and dips 80° SE. The zone of mineralization ranges from 5 to 50 feet in width, but the average is 12 to 15 feet. On the footwall side of the ore zone is a dike of light-colored diorite 1 to 2 feet in width. Also present in the ore zone are a number of narrow basic dikes.

The mine is developed by three shafts, the Eureka, old Beebe, and Beebe No. 2 shafts. Levels were driven at 130, 250, 370, 500, 600, and 700 feet. On the 370-foot level, a length of 700 feet was drifted in ore (Logan, 1938, p. 217). A winze was sunk from the 500-foot level to the 700-foot level. The last production in the mine was from stopes driven from the 600- and 700-foot levels. Shrinkage stopes were employed.

Ore from this mine and the Alpine mine was treated in the mill, which could handle up to 250 tons per day (Logan, 1938, p. 218).

Flotation was employed, and the flotation concentrates were treated by cyanidation (Logan, 1934, p. 209). The mill originally was equipped with two Hadsell mills which later were replaced by conical ball mills. An average of 20 men worked at the mine and mill.

Big Canyon (Oro Fino) Mine. Location: sec. 29, T. 9 N., R. 10 E., M. D., 4½ miles south of Shingle Springs in Big Canyon. Ownership: Capitol Company, 1 Powell Street, San Francisco.

One of the larger gold producers in the county, the Big Canyon mine, originally was active prior to 1888 (Irelan, 1888, p. 174). Between 1893 and 1901, the mine produced $720,000 (Logan, 1938, p. 220). The property was idle until around 1915, when some development work was done (Tucker, 1919, p. 293). In 1934, the property was acquired by the Mountain Copper Company. This concern sank a new shaft, erected a mill, and operated the mine on a large scale until 1940. The total gold output during this last operation was $2,368,000 (Mountain Copper Company, personal communication, 1956). The mine has been idle since 1940.

The deposit which is in the West gold belt, is in a shear zone with amphibolite to the west and a serpentine lens to the east. East of the serpentine are metasedimentary rocks of the Calaveras group. The ore body is a vein-like mass of sheared and brecciated meta-andesite containing free gold, pyrite, pyrrhotite, albite, quartz, ankerite, and arsenopyrite. It merges into the amphibolite. The gold is in the quartz, pyrite, and pyrrhotite. The best ore was found to consist of equal parts albite, ankerite, and quartz, of which the albite appeared to be highly favorable for gold deposition (Logan, 1938, p. 220). Free gold constituted about 20 percent of the total gold contained, the remainder being combined with pyrite and pyrrhotite (Huttl, 1935, p. 217). Ore yielding $5 or more per ton was considered satisfactory.

The ore body strikes north to N. 25° E. and dips 35° to 40° E. and SE. The ore shoot that was mined during the 1930's had a maximum length of 450 feet and was as much as 60 feet wide.

The mine originally was developed by a vertical shaft sunk to a depth of 200 feet and then sunk an additional 540 feet on a 40° incline to the east. The ore was stoped out to the surface from the 500-foot level of this shaft during the early operations (Tucker, 1919, p. 293). The Mountain Copper Company sank a 620-foot shaft on a 45° incline, 400 feet north of the old shaft. Drifts were extended several thousand feet along the strike of the ore body. The ore was mined in open stopes. In 1937, some ore in the upper portion of the mine was mined by open pit methods.

From the mine, the ore was belt-conveyed to the mill. The mill had a capacity of 300 tons (Huttl, 1935, p. 217). The mill was equipped with two 7- by 6-foot ball mills, two Dorr classifiers, and five Fagergren flotation cells and a cleaner cell. Concentrate was sent through a Dorr thickener and an Oliver filter. Concentrates were shipped to the smelter at Tacoma. When in full operation, about 300 tons of ore per day were mined. A crew of 150 men worked at the mine and mill.

Big Sandy (James Marshall) Mine. Location: sec. 24, T. 11 N., R. 10 E., M. D., on the Mother Lode half a mile south of Kelsey. Ownership: H. J. Picchetti, 1115 Fairview Avenue, San Jose.

This gold mine is of historical interest as it was located originally by James W. Marshall (Logan, 1934, p. 19). It was active during the 1890's, and the ore was treated in a 10-stamp mill (Crawford, 1896, p. 134). During the 1930's the mine was worked for pocket gold, and several fine specimens of crystallized gold were found.

The deposit is on the Mother Lode belt in Mariposa slate. The vein, which is as much as 15 feet in width, consists of bands of quartz and ankerite. Associated with the vein are amphibolite schist, talc schist, and dike rocks. The vein strikes north-northwest and dips east. Most of the ore produced from the mine was low-grade with average values ranging from $1.75 to $2.25 per ton (Logan, 1934, p. 20). The specimen ore was obtained largely from seams in an altered dike.

The low-grade ore was mined from an open cut 750 feet long. The mine also is developed by a 340-foot vertical shaft with levels at 60, 120, 227, and 323 feet with crosscuts to the vein. The vein is cut by the shaft at a depth of 240 feet.

Black Oak (Clark, Davey, Dayton Consolidated) Mine. Location: sec. 34, T. 12 N., R. 10 E., M. D., at Garden Valley just south of the school. Ownership: R. J. Wilson, Garden Valley.

The Black Oak mine was worked originally as a pocket gold mine prior to 1934 (Logan, 1934, p. 20). In 1934, the mine was reopened by the present owner. A new shaft was sunk, a mill erected, and the property soon developed into one of the more important sources of gold in the county. Although the deposit is relatively shallow, an extensive amount of crosscutting was done. By 1937, more than $400,000 had been produced besides a considerable amount of gold that was believed to have been stolen by "high-graders" (Logan, 1938, p. 224). Operations continued until 1942, when the mine was shut down. It has been idle since. The total output of the mine is approximately $1,250,000 (F. H. Frederick, personal communication, 1955).

Between 1934 and 1938, the Dayton Consolidated Mines Corporation developed the Davey claim, which adjoins the Black Oak mine on the north and northeast, the Clark claim on the east, and the Davenport claim on the southeast. The Dayton shaft or winze was sunk on the Davey claim, and some ore was produced. These properties were merged with the Black Oak mine in 1938 (Logan, 1938, p. 224) and now are considered to be part of the Black Oak mine.

This deposit is just north of where the Mother Lode belt splits into two branches—one branch extending northwest through Greenwood, the other north through Georgetown. The structure of the general area of this deposit is characterized by tightly compressed, steeply plunging folds. Thrust faults dipping 20° to 35° E. have disrupted the contacts.

The Black Oak vein zone strikes northwest and dips steeply northeast. It lies in the mass of amphibolite schist that is between the two belts of Mariposa slate in this area. However, there is a strip of graphitic slate 50 feet wide that forms the hanging wall. The main ore shoot, which consisted of gold-bearing stringers and veins of calcite and quartz, was wide at the north end and pinched out to the south. It averaged 110 feet in length, had a vertical extent of 380 feet, and an average width of 12 feet; it pitched 80° S. The shoot followed or was near the contact between the graphitic slate and amphibolite. In the

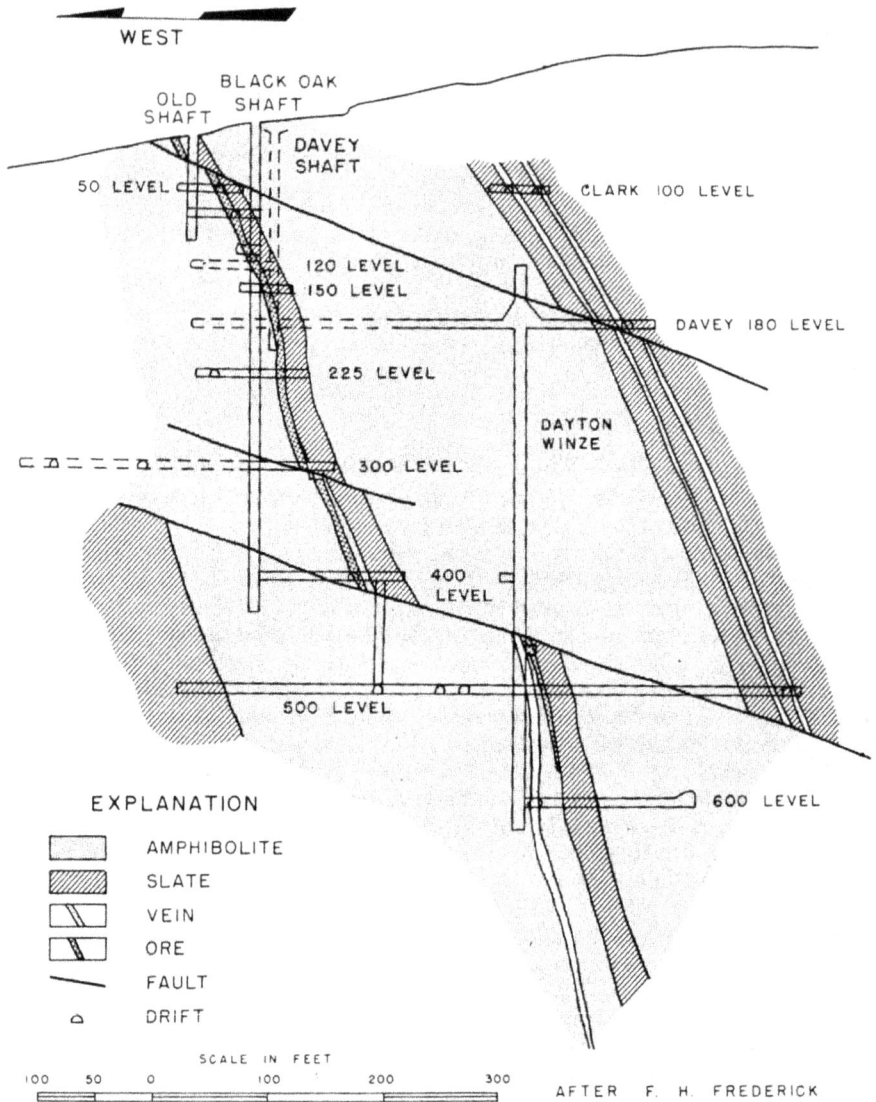

FIGURE 17. Cross-section of underground workings of Black Oak gold mine, showing veins.

upper levels of the Black Oak mine, it followed the contact and in the lower levels it was enclosed in amphibolite but near the slate. The vein began to veer away from the contact below the 500-foot level, and little ore was found below this level (see Fig. 17). Apparently, gold deposition was controlled by the slate (F. H. Frederick, personal communication, 1955). A number of low-angle thrust faults offset the vein and in some cases localized the gold mineralization. The principal fault offset the vein and ore shoot for about 200 feet between the 400- and 500-foot levels. Only about $100,000 was produced from the ore shoot below this fault (F. H. Frederick, personal communication, 1955).

About 250 feet east of the Black Oak vein is the Clark vein zone, which is in slate. To the north the veins converge with the slate-amphibolite contact. The vein matter consists of quartz with feldspar, calcite, and appreciable amounts of pyrite. Pyrite also is present in the slate wall rock. About 100 feet east of the Clark vein is the Davenport vein zone, from which there was a small amount of ore produced. These two veins also are cut by the thrust faults.

The main working entry is the 400-foot vertical Black Oak shaft. The Dayton shaft, which actually is a vertical winze, was sunk from the 175- or the Davey 180-foot level to the 600-foot level 225 feet east of the Black Oak shaft. There are about 6000 feet of drifts and cross-cuts. Crosscuts extend east from the Dayton shaft to the Clark vein zone, but there has been very little stoping in this portion of the mine. Other shafts on the property include the 180-foot Davey shaft to the north and the 100-foot Clark shaft to the south. A 180-foot crosscut extends east from the Clark shaft to the Clark vein zone.

During the later operations the ore was treated in a 35-ton mill which included primary crushers, Williamson ball mill, Dorr classifier, a hydraulic trap later replaced by a Denver jig, and three Fagergren flotation cells.

Blue Gouge (Berg) Mine. Location: sec. 21, T. 10 N., R. 13 E., M. D., by Camp Creek 6 miles north-northwest of Grizzly Flat. Owner-ship: Americo and Columbus Sciaroni, Grizzly Flat.

In 1896, the Blue Gouge gold mine was extensively prospected by Mackay, Flood, and associates of San Francisco (Crawford, 1896, p. 135). However, this work was abandoned soon afterward (Logan, 1938, p. 225). Later some work was done on the property by the present owners, ending about 1925 (Americo Sciaroni, personal communication, 1955). A small amount of work was done again about 1936 (Logan, 1938, p. 225).

The deposit consists of a zone of parallel gold-bearing quartz veins in slate and chlorite schist. Some of the slate and schist was gold bearing. The ore zone is as much as 400 feet wide and is exposed for a length of 3500 feet. It strikes northwest and dips northeast. The quartz veins range from 6 to 16 feet in width. The footwall is slate, the hanging wall granodiorite. The grade of ore in the principal veins averaged $7 per ton in gold at the old price, and the mineralized slate and schist was somewhat less valuable (Crawford, 1896, p. 135). The mine is developed by seven crosscut adits ranging from 120 to more than 300 feet in length.

Briarcliffe (Baldwin, Last Chance, National) Mine. Location: secs. 1, 2, 11, and 12, T. 8 N., R. 10 E., M. D., 1 mile southeast of Nashville. Ownership: Nick Neilsen, Box 100, Diamond Springs.

The Briarcliffe gold mine, which consists of the Last Chance claim on the north and the Baldwin claim on the south was active in the 1890's (Crawford, 1894, p. 102). The Last Chance claim again was active in 1909, when a 10-stamp mill was in operation (Logan, 1938, p. 226). The Baldwin claim was reopened about 1914, and some gold was produced (Tucker, 1919, p. 280). In 1932, these properties were obtained by Briarcliffe Mines, Ltd. and operated from 1932 to 1936 and again during 1940-41. Most of this later work was in the Baldwin claim.

Although the deposit is a mile east of the Mother Lode, its veins and ores are similar. The vein, which strikes N. 10° E. and dips 60° to 72° SE., has an amphibolite hanging wall and slate footwall. The average width of the vein is 15 feet. The best ore was on the hanging wall side of the vein and extended into the amphibolite. Up to 1936, more than 30,000 tons of ore that yielded $3 to $4 per ton had been mined (Logan, 1938, p. 226).

The mine is developed by a 200-foot inclined shaft, a northeast drift adit that is connected with the shaft, and a 500-foot winze sunk from the adit about 500 feet in from the portal. Most of the ore mined in later operations was from the area of the winze. The ore was treated by flotation in a 100-ton mill.

Church Mine. Location: sec. 12, T. 9 N., R. 10 E., M. D., on the Mother Lode belt 2 miles southeast of El Dorado by Deadman Creek. Ownership: Madre de Oro Gold Mines Company, Box 925, Corcoran.

One of the better-known Mother Lode gold mines in El Dorado County, the Church mine was first worked on a small scale about 1850 (DeGroot, 1890, p. 171). During the 1860's, it was consolidated with the Union mine, which lies to the south and was later worked separately (see Union mine). By 1868, these properties had produced more than $600,000 (Logan, 1934, p. 42). The mine was worked extensively during the 1880's and 1890's. By 1896, it had been developed to a depth of 1200 feet, and the ore was treated in a 10-stamp mill (Crawford, 1896, p. 137). In 1900, the mine was 1350 feet deep (Storms, 1900, p. 92). In 1907, it was shut down (Logan, 1926, p. 413).

In 1941, the mine was reopened by the Madre de Oro Mining Company. The shaft was rehabilitated, a new surface plant built, and a 20-stamp mill with flotation erected. Production figures of the U. S. Bureau of Mines Minerals Yearbook for 1941 show that 496 tons of ore were milled, from which 169 ounces of gold were recovered. The mine was shut down in 1942 and has been idle since.

In the spring and summer of 1953, the 20-stamp mill was used as a custom chrome mill by El Dorado Chrome Company (see El Dorado Chrome Company in Chromite section). Late in 1953 and early 1954, the mill was used by the North American Tungsten Company to treat tungsten ore that was trucked in from the Garnet Hill mine in eastern Calaveras County.

The three main parallel veins are enclosed in Mariposa slate. They range from 5 to 10 feet in thickness, strike north to northeast, and dip steeply east or southeast. The principal work and largest amount of production was from the middle or Kidney vein. The west vein contains only low-grade material. Some work was done on the east vein in the Union mine, which is just south of the Church mine.

The Kidney vein strikes north between the surface and the 350-foot level. Below this level the strike swings to the northeast. From the 350- and 500-foot levels, the vein dips 45° E., but below the 500-foot level it steepens to 74° SE. A number of kidney-shaped ore shoots were developed between the surface and 500-foot level. The main ore shoot was continuous between the 500- and 1200-foot levels; below the 1200-foot level, the grade decreased greatly (Tucker, 1919, p. 283).

The best ore, as high in grade as $30 per ton (Storms, 1900, p. 92), was mined during the earlier operations. Ore milled from as far down as the 1300-foot level had averaged $17 per ton (Logan, 1934, p. 22). However, below that level it decreased to $4 per ton or less. Sulfides formed 1½ to 2¼ percent of the ore and assayed as high as $140 per ton in gold (Logan, 1934, p. 22). Considerable amounts of gouge were present, especially in the lower levels. Some of the gouge carried some gold.

The mine is developed by a 1200-foot vertical shaft, with crosscuts to the vein. Levels are at 100, 200, 300, 350, 500, 600, 700, 850, 1000, and 1200 feet. On the 1200-foot level, a 670-foot crosscut extends east to the vein, from which a winze was sunk on the vein to the 1400-foot level. On the 600- and 1200-foot levels, drifts extend north and south to the property line.

There is an extensive surface plant including a machine shop, living quarters, a warehouse, and a conveyor belt running from the bins on the steel headframe to the mill. The 20-stamp mill is equipped with Deister tables and a flotation circuit.

Cosumnes (Melton, Middle End) Mine. Location: secs. 3 and 4, T. 9 N., R. 13 E., M. D., 2½ miles north of Grizzly Flat. Ownership: Caldor Lumber Company, Diamond Springs.

The Cosumnes gold mine was active during the 1880's, and the ore was treated in a 15-stamp mill (Irelan, 1888, p. 179). In these early operations, it was known as the Melton mine. The property was active again in 1894 (Crawford, 1894, p. 117). The mine was reopened in 1928 as the Middle End mine (Logan, 1938, p. 238) and was operated almost continuously until 1942. In 1938, Cosumnes Mines, Inc. was organized. The mine was reopened in 1945 and operated by this concern until 1950 when it was purchased by the present owner.

Most of the ore was mined from the main or Middle End vein, which strikes north-northeast and dips steeply west. The vein has an average width of 3 feet. Country rock is granodiorite. A number of ore shoots were worked. Appreciable amounts of galena and pyrite with sphaleriate, chalcopyrite, stibnite, and arsenopyrite are present in the ore. During the last years of operation, the ore contained as much as $25 per ton in gold (Columbus Sciaroni, personal communication, 1955). Sulfide concentrates yielded from $100 to $200 per ton in gold (Logan, 1938, p. 238).

The mine is developed by a 380-foot southwest crosscut adit to the vein and several thousand feet of drifts. The ore was treated in a 60-ton mill equipped with amalgamation and flotation units and the flotation concentrates trucked to a smelter.

Crystal (El Dorado Crystal) Mine. Location: sec. 18, T. 9 N., R. 10 E., M. D., 3 miles south of Shingle Springs and half a mile south of Frenchtown. Ownership: F. W. Barrette, Shingle Springs.

This gold mine was originally worked prior to 1890. During the early 1890's, the mine was worked through a 250-foot inclined shaft and 350-foot crosscut adit. The ore was treated in a 10-stamp mill (Crawford, 1894, p. 108). After being idle for many years, the mine was reopened in 1937 by Ben Lockwood of Shingle Springs (Logan, 1938, p. 228). Also a new mill was erected. The mine was operated until 1940 and has been idle since.

The deposit lies in the West Gold Belt. It consists of an irregular, north-trending quartz vein as much as 12 feet wide. Country rock is amphibolite-chlorite schist. Gabbrodiorite lies to the west. Apparently the vein material was deposited in irregular cracks and occasional cavities, which resulted from the intrusion of the gabbrodiorite into the schist (Logan, 1938, p. 228). There are a number of small open cavities in the vein in which euhedral quartz crystals have grown. The vein matter consists of quartz, feldspar, calcite, and pyrite. The ore contained as much as $11 per ton in gold (Logan, 1938, p. 229).

The mine is developed by a 250-foot inclined shaft and a 1,028-foot adit driven north. A 480-foot inclined winze with levels at 200, 326, and 456 feet was sunk from the adit 328 feet in from the portal. An ore shoot ranging from 8 to 24 feet in width and up to 280 feet long was developed on the 326-foot level.

The mill had a capacity of 125 tons of ore per day. It was equipped with a ball mill and a cyanidation unit.

French Creek Mine. Location: sec. 31, T. 9 N., R. 10 E., M. D., 3½ miles northeast of Latrobe. Ownership: William Lange and L. W. Loomis, Placerville.

This gold property has been under development since 1953. The operators are sinking an inclined shaft, which was 30 feet deep early in 1956. A surface plant, including a headframe and milling equipment, is being installed.

The deposit, which is in a belt of metavolcanic rock west of the Mother Lode gold belt, is several hundred yards south of the old Brandon mine. The ore body is dark green, fine- to medium-grained metadiabase with disseminated auriferous pyrite and some fine free gold. Minor amounts of chalcopyrite and a trace of arsenopyrite are present.

Strike of the ore body is north, and dip is 65 to 68° E. The ore body has been traced for about 300 feet along the strike, and its average width is 20 feet. Assays range from $2.80 to $60 per ton in gold. The estimated average is $20 to $30 (William Lange, personal communication, 1954). The shaft is sunk in the footwall of the ore body.

Funny Bug (Pendelco) Mine. Location: sec. 5, T. 10 N., R. 10 E., M. D., 1 mile southwest of Gold Hill on the north bank of Webber Creek. Ownership: R. V. Montgomery, Route 2, Box 193-K, Placerville.

This mine was active intermittently from 1928 to 1942 and yielded small amounts of gold and copper. In 1953, the property was leased by Carl Howe of Placerville who did a small amount of rehabilitation work and surface exploration.

Two northeast-striking veins lie in a contact metamorphic zone of greenstone, mica schist, and granodiorite. The ore consists of parallel bands of pyrite and chalcopyrite associated with quartz and magnetite. Bornite, specular hematite, sphalerite, molybdenite, arsenopyrite, stibnite, and galena are also present (Logan, 1938, p. 242). The mine is developed by crosscuts on two levels from a 200-foot shaft sunk near the main vein.

Grand Victory Mine. Location: secs. 33 and 34, T. 10 N., R. 11 E., M. D., on Squaw Creek 4 miles southeast of Diamond Springs. Ownership: Provident Minerals Corporation, c/o Floyd Singleton, 459 Turk Street, San Francisco 2.

This gold mine first was worked in 1857, and in 1879 the ore was treated in a five-stamp mill (Logan, 1938, p. 231). By 1888 extensive open cuts and underground workings had been developed, and the mill had been enlarged to 40 stamps (Irelan, 1888, p. 194). In 1894, a cyanidation plant was installed and was operated until 1901 when the mine was shut down. During the 1930's, a considerable amount of prospecting and sampling was done on the property, and a number of drifts and crosscuts were driven.

The deposit is a few miles east of the Mother Lode. Country rock is silicified schist, slate, quartzite, and limestone of the Calaveras group. The ore consists of silicified schist and slate with quartz seams containing free gold, pyrite, and smaller amounts of arsenopyrite. The strike of the ore bodies varies from northwest to northeast, and in some places it cuts the schistosity of the enclosing rock. The ore bodies are as wide as 100 feet (Crawford, 1894, p. 112). Much of the ore mined during the early operations was highly weathered material. There are a number of diorite dikes parallel to the ore bodies.

The mine is developed by a 500-foot drift adit and open cuts as wide as 135 feet, which extend over 500 feet along the strike. About 450 feet in from the adit portal is a winze with levels at 100, 200, and 300 feet. There are several thousand feet of drifts and crosscuts.

Grit (Liddicoat, Spanish Dry Diggings) Mine. Location: sec. 29, T. 13 N., R. 10 E., M. D., at Spanish Dry Diggings, 4 miles north of Greenwood. Ownership: Liddicoat Gold Mines, Inc., Route A, Box 27, Greenwood; J. L. Liddicoat, president.

This gold mine is at the north end of the west branch of the Mother Lode. The mine was highly productive around 1852 and between 1860 and 1867, yielding $100,000 (Logan, 1934, p. 46). In 1865, a mass of crystallized gold weighing 101.4 ounces Troy was found on the property. The mine was active again from 1919 to 1922 (Logan, 1922, p. 44). The total output of the property up to this time was about $300,-000. The present owners operated the mine almost continuously from 1945 to 1952 and produced a total of $34,000 (J. L. Liddicoat, personal communication, 1955).

The deposit consists of gold-bearing quartz and calcite veins and veinlets in amphibolite schist and Mariposa slate. The upper portion of the deposit is highly decomposed and was worked by hydraulicking to depths of 125 to 150 feet in the early days. Such ground contained as much as $2 to $3 per cubic yard (Logan, 1920, p. 426). Three parallel veins composed of quartz and calcite strike N. 20° W., dip 80° NE., and have been explored by underground workings. The veins pinch and swell abruptly and range from a few feet to as much as 20 feet in thickness. Considerable amounts of coarse crystalline pyrite are present. Small patches of serpentine are exposed west of the west vein.

All of the ore mined since 1945 has been from the west vein which is developed by an 800-foot southeast drift adit. About 400 feet in from the adit portal, crosscuts were driven 70-feet west and 40-feet east. The vein has been stoped for a distance of 125 feet, beginning about 275 feet in from the adit portal. This stope raises into the open cut that was hydraulicked in the early days. The ore was trammed out the adit to the mill which lies just north of the portal.

GOLD ORE

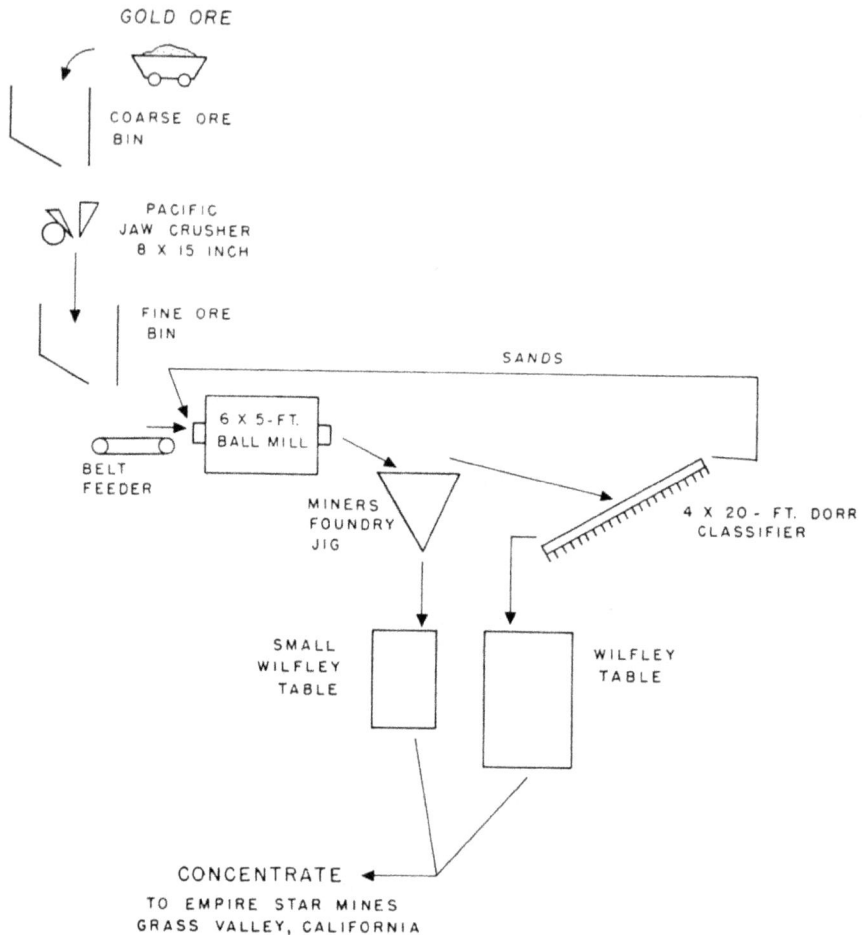

COARSE ORE
BIN

PACIFIC
JAW CRUSHER
8 X 15 INCH

FINE ORE
BIN

SANDS

6 X 5-FT.
BALL MILL

BELT
FEEDER

MINERS
FOUNDRY
JIG

4 X 20- FT. DORR
CLASSIFIER

SMALL
WILFLEY
TABLE

WILFLEY
TABLE

CONCENTRATE
TO EMPIRE STAR MINES
GRASS VALLEY, CALIFORNIA

FIGURE 18. Mill flowsheet of Grit gold mine.

At the mill, which has a capacity of 80 tons, the ore was concentrated by jigging and tabling. Concentrates were sold to the Empire-Star mine at Grass Valley. The mill heads averaged $18 per ton in 1952 (J. L. Liddicoat, personal communication, 1953).

Hazel Creek Mine. Location: sec. 3, T. 10 N., R. 13 E., M. D., 2 miles southeast of Pacific House on Hazel Creek and 15 miles east of Placerville. Ownership: Fay. M. R. Gunby, Box 752, Placerville, leased by Hazel Creek Mining Corporation, P. O. Box 508, North Sacramento, Richard Ronne, president and Ray Graetz, superintendent.

The Hazel Creek mine has been active continuously since 1948 and is the largest gold mining operation in El Dorado County. In 1948 the vein was discovered in a logging road cut (Earl Estey, former superintendent, personal communication, 1954). Soon afterward, an adit was driven and a mill erected. At the present time, 30 to 40 tons of ore per day are being mined and milled.

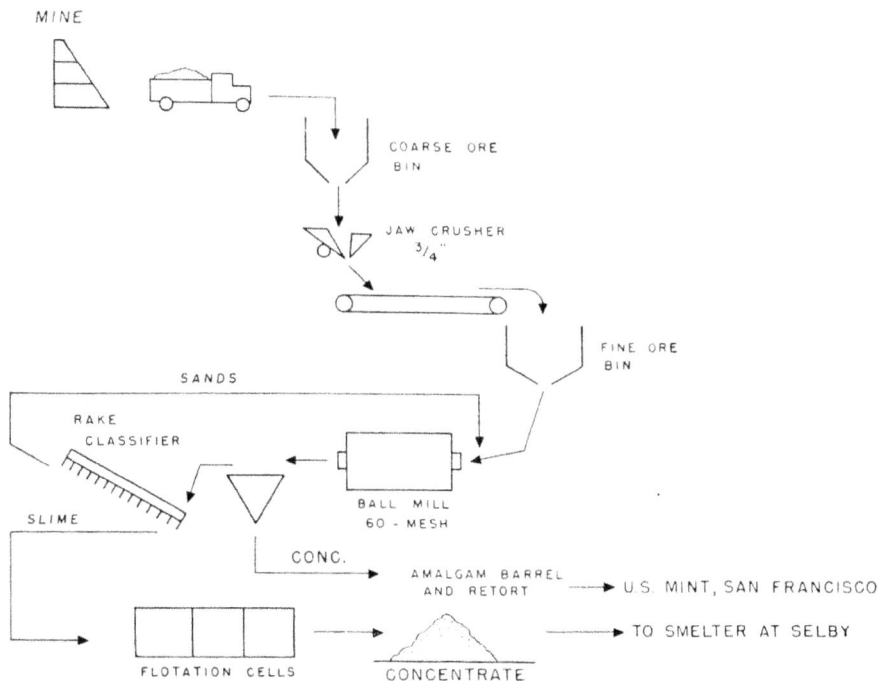

FIGURE 19.   Mill flowsheet, Hazel Creek gold mine,
Hazel Creek Gold Mining Company.

This is an East Belt gold deposit. There are two principal gold-bearing quartz veins that are nearly parallel and about 40 feet apart. They have a north-northwest strike. The east vein has an average dip of about 85° E. and joins the vertical west vein above the adit portal. The west vein is 5 to 6 feet wide, while the east vein ranges from 5 to 12 feet in width. There are numerous quartz stringers parallel to the principal veins especially in the south portion of the deposit. About 300 feet south of the adit portal, a southeast-striking cross vein several feet wide extends from the east to the west vein. In the north portion of the mine, the east vein enters a shear zone where there are considerable amounts of gouge. Here the strike swings to the east, and the width of the vein narrows to less than a foot. Country rock is interbedded slate and green schist with smaller amounts of quartzite and metachert, all of which are part of the Calaveras group. A number of post-mineral dikes of fine- to medium-grade diorite and quartz diorite cut both the veins and country rock.

The ore consists of native gold with considerable amounts of galena and pyrite in quartz, and smaller amounts of chalcopyrite and sphalerite. The free gold usually is associated with the galena. Considerable amounts of high-grade ore were produced from the area where the two main veins join in the upper level and where the cross vein intersects the east vein. (Ray Graetz, personal communication, 1955).

The mine is developed by a 500-foot drift adit driven south on the west vein and a 130-foot inclined two-compartment shaft sunk on the west vein near the adit portal. The 100-foot level, which is now 700

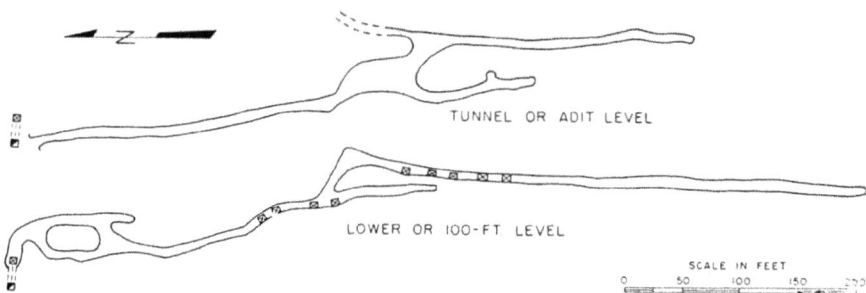

FIGURE 20.   Map of underground workings of Hazel Creek gold mine.

feet long, is connected with the adit level by raises and stopes. The veins are mined by shrinkage stopes. Lightweight Swedish Atlas Copco drills with tungsten carbide inserts are used in drilling. Seven-foot holes are used for development headings and 5-foot holes are used in the stopes. Forty percent gelatin dynamite and electric caps are used for blasting. Early in 1956, most of the ore was being produced from an extensive stope in the south portion of the 100-foot level on the west vein. This stope is about 300 feet long, 50 feet high, and averages 10 feet in width.

Ore from the mine is treated at the mill which is on the north bank of Hazel Creek. The mill, which employs flotation, has a capacity of 30 tons per day. The free gold is amalgamated and retorted, and the sponge gold is sent to the smelter at Selby. The sulfied concentrate from the flotation cells is dried and also sent to Selby. The sulfide concentrate may assay as high as $500 per ton in gold and silver (Earl Estey, personal communication, 1954). From 10 to 15 men work at the mine and mill.

Kelsey (Lady) Mine.   Location: secs. 24 and 25, T. 11 N., R. 10 E., M. D., half a mile southeast of Kelsey. Ownership: A. D. O. Crabtree, 616 North F Street, Porterville.

The Kelsey gold mine was worked originally prior to 1915 (Tucker, 1919, p. 289). In 1926, the Kelsey Mining Company was formed which reopened the mine and erected a mill (Logan, 1934, p. 29). It was active until 1931. The mine was reopened again in 1934, the mill altered, and was operated until 1941.

The deposit is in Mariposa slate associated with narrow bands of amphibolite and serpentine. The ore consists of schist and slate with quartz containing free gold, pyrite, and some galena. Considerable amounts of ankerite are present. The rocks in this area trend northwest. The formations are much disturbed by faulting. The ore mined during 1928-31 yielded from $1.80 to $6.40 per ton (Logan, 1934, p. 29), while much of the ore mined later on was even lower in value (Logan, 1938, p. 235).

The mine is developed by a main 1700-foot north drift adit and a 700-foot north drift adit about 300 feet above the main adit. These two adits are connected by a raise. Also there is a 42-foot shaft near the main adit portal.

The mill first erected by the Kelsey Mining Company had ten stamps, a Williamson ball mill, classifier, four Kraut flotation cells, and a cleaner cell. Later the stamps were replaced by a Telsmith gyratory crusher and Aurora jaw crusher. Also a Pan-American jig was added. About 40 tons of ore per day were milled.

Larkin (Diamond Springs) Mine. Location: sec. 29, T. 10 N., R. 11 E., M. D., on the Mother Lode 1 mile east of Diamond Springs. Ownership: Bernard B. Ball, 122 Bedford Avenue, Placerville.

The Larkin gold mine also has been a source of copper ore. It is noted for the large dolomitic vein on the property. The mine was active in 1896 when it was developed by a 250-foot vertical shaft (Crawford, 1896, p. 148). The ore was treated in a five-stamp mill. By 1900 the shaft had been extended to a depth of 600 feet, and the mill had been increased to 10 stamps (Storms, 1900, p. 93). The shaft later was extended to 800 feet in depth. In 1903, the mine was shut down. It was reopened in 1918, and some copper ore was produced (Logan, 1926, p. 407). The total output of the mine is estimated to be $125,000 (Logan, 1934, p. 30).

Several gold-bearing quartz veins are in the hanging wall side of a large mass of dolomite. This large mass of dolomite material, which is as much as 80 feet wide, contains iron oxide, and has also been classed as ankerite (Storms, 1900, p. 93). It is partially altered to talc schist, and is cut by dikes of diabase. Country rock on both sides of it is Mariposa slate. The veins have a north-northeast trend. The principal productive vein ranged from 4 to 12 feet in width and contained 1½ percent pyrite (Logan, 1934, p. 30). Chalcopyrite and malachite are also present. Several ore shoots that contained as much as 10 percent copper were developed (Aubury, 1908, p. 217). A west vein, about 7 feet wide, was prospected on the 400-foot level.

Martinez (Hillside Group) Mine. Location: secs. 12 and 13, T. 9 N., R. 10 E., M. D., on Martinez Creek south of the Union mine about 4½ miles southeast of El Dorado. Ownership: Martinez Gold Mines Company, P. O. Box 61, El Dorado.

In 1915, the Martinez gold mine was operated by the Hillside Gold Mining Company (Tucker, 1919, p. 288). It was worked on a small scale around 1926 (Logan, 1926, p. 415) and again during the early 1930's. In 1937, some development work was done in the lower workings of the mine, and a small amount of ore was produced (Logan, 1938, p. 237).

This deposit consists of a series of parallel veins on the east side of the belt of Mariposa slate that contains the Mother Lode in this area. The veins are in slate and at the contact with the Calaveras formation that lies to the east. The veins strike N. 20° E. and dip 70° SE. Interbedded layers of greenstone are associated with the veins; mariposite and ankerite are also present. The ore mined in 1917 yielded $4.12 per ton (Logan, 1938, p. 412).

The mine is developed by a 600-foot crosscut adit driven west with numerous raises and drifts. South of this is a 600-foot crosscut adit and on the hill above are a number of older workings. The ore was treated in a five-stamp mill.

Monarch-Sugar Loaf Mine. Location: sec. 27, T. 9 N., R. 10 E., M. D., 3 miles northwest of Nashville. Ownership: leased by George Ross, Placerville.

The Monarch-Sugar Loaf gold mine was worked originally in the 1850's (George Ross, personal communication, 1955). The mine was active between 1870 and 1907, and the ore was treated in a 10-stamp mill. During the 1930's, some prospecting was done on the property. In 1953-54, the mine was leased by J. H. Wren and associates of Sacramento who rehabilitated some of the workings. Early in 1955, George Ross and associates leased the mine; they since have mined some high-grade ore from small pockets in the vein.

The quartz vein is in fine-grained meta-andesite of the Logtown Ridge formation (upper Jurassic). It has a north to northwest strike and a flat dip to the west and southwest. Near the surface, the gold is chiefly in small but rich ore shoots with little or no sulfides. Much of the recent output has been specimen material. The amount of sulfides present in the ore increased at depth (George Ross, personal communication, 1955). Chalcopyrite and sphalerite are the most common sulfides; pyrite and galena ore present in smaller amounts. The most favorable areas for the occurrence of gold in the vein are where it swells, changes in strike or dip, and junctions with small stringers.

The deposit has been developed by open cuts and shallow shafts for about 2,000 feet along the strike. Present work is confined to a 70-foot crosscut adit, and several nearby open cuts in the central portion of the deposit. The ore is treated by hand-sorting, hand-mortaring, and amalgamation. Three men work intermittently at the mine.

Montezuma-Apex and Montezuma Extension Mine. Location: sec. 35, T. 9 N., R. 10 E., and sec. 2, T. 8 N., R. 10 E., M. D., just east of the Cosumnes River. Ownership: Montezuma-Apex Mining Company, c/o National Tunnel and Mines Company, 818 Kearns Building, Salt Lake City, Utah.

Originally worked at shallow depths during the early days of the gold rush, the Montezuma mine was active up to 1871 (Logan, 1934, p. 31). The mine was worked from 1890 to 1907 and again during 1914. From 1920 to 1928, there was some recorded output. In 1931, the Nashville Mines, Ltd. reopened the mine, and in 1933, they were succeeded by the Montezuma-Apex Mining Company which operated the property until 1939. This concern also operated the Havilah (Nashville) mine which lies just to the south.

The vein occurs in the Mother Lode belt and is largely in Mariposa slate. Small bands of greenstone are interbedded with the slate, and some parts of the hanging wall contain schist. Also there is a heavy gouge in portions of the footwall. The vein strikes north to northeast and has an average dip of 60° E. The mine is developed by a 1,540-foot inclined shaft and an older 360-foot inclined shaft 300 feet to the north. In the old workings an ore body 8 to 20 feet wide was stoped for a length of 250-feet down to the 120-foot level (Logan, 1934, p. 31).

In the newer workings, an ore shoot 150 feet long was mined north of the shaft between the 800- and 1,000-foot levels in 1914. When the mine was reopened in 1932, these levels again were worked. The shaft was then deepened, and a new ore shoot that was about 570 feet long and 4 to 6 feet wide was developed on the 1,200-foot level. Most of the later output was from this shoot. Also, a spur vein with a northwest strike entered the main vein from the east side near the shaft and was mined between the 900- and 1,100-foot levels. Below a depth of 1,225 feet, however, the ore decreased in grade (Logan, 1938, p. 239). The 1,500-foot level was extended 1,050 feet south under the Nashville mine, which is 1,000 feet to the south, with the object of encountering the Havilah (Nashville) vein.

Ore from the mine originally was treated in a 10-stamp mill. This was replaced in 1933 by a 240-ton mill equipped with two Marcy ball mills, hydraulic traps, Wilfley tables and a 10-cell flotation unit (Logan, 1934, p. 206).

Mount Pleasant Mine. Location: sec. 16, T. 9 N., R. 13 E., M. D., 1 mile west of Grizzly Flat. Ownership: Annie S. Kirk, Box 507, Placerville.

The principal source of gold of the Grizzly Flat district, the Mount Pleasant mine, first was worked in 1851, although the output at that time is unknown (Logan, 1926, p. 416). From 1874 to 1914, the mine yielded $1,046,748 in gold (Logan, 1926, p. 416). After 1914 there has been only a small amount of prospecting on the property. It was last worked by E. W. Morey of Grizzly Flat during the period of 1939-41.

The deposit consists of a belt of nearly parallel quartz veins in a contact zone between granodiorite on the west and mica schist on the east. The belt is 300 feet wide and strikes N. 13° E., the three main veins in the belt strike from 2° to 10° east of the strike of the belt in which they lie. There are numerous spurs and lesser veins parallel to the principal veins. The quartz bodies of the veins are lenticular (Tucker, 1919, p. 292).

The ore shoots consist of lenticular bodies of banded quartz containing free gold, pyrite, galena, and sphalerite, and some chalcopyrite. The grade of the ore mined during the later operations is unknown, but between 1881 and 1887 it averaged $14 per ton (Logan, 1926, p. 416).

Most of the ore mined has been from the Earle vein on which most of the development work has been done (Logan, 1938, p. 241). The vein is developed by a 1,065-foot shaft with levels at 100-foot intervals. Most of the development was north of this shaft, and much of the output was from above the 850-foot depth. Drifts on the various levels range from 300 to 1,300 feet in length. There are more than 9,000 feet of drifts. Also there are two other shafts 300 to 600 feet deep. The ore was treated in a 10-stamp mill that burned down prior to 1926 (Logan, 1926, p. 417).

Nashville (Havilah, Tennessee-Nashville) Mine. Location: sec. 2, T. 8 N., R. 10 E., M. D., on the Mother Lode at Nashville. Ownership: Montezuma-Apex Mining Company, c/o National Tunnel and Mines Company, 818 Kearns Building, Salt Lake City, Utah.

The Nashville gold mine was worked first in 1851 and was credited with an output of $150,000 from shallow workings (Logan, 1934, p. 27). Much development work was done during the period of 1868-71 and again around 1880. A 20-stamp mill was erected on the property in 1894. The mine was active again from 1903 to 1906. The Montezuma-Apex Mining Company operated the mine from 1934 to 1936, and the ore was treated at the Montezuma mill. The mine has had a total output valued at about $400,000 (Logan, 1938, p. 233).

The vein, ranging from 5 to 20 feet in width, is in Mariposa slate. It strikes N. 10° E. and dips 60° SE. The ore was free milling containing 2 percent pyrite (Tucker, 1919, p. 292). Some of the ore produced during the 1930's averaged $5 per ton (Logan, 1938, p. 233).

The mine is developed by a 1200-foot inclined shaft with levels at each 100 feet. It is connected with the Montezuma mine to the north on the 1200-foot level. The last work done in the mine was on the 200-foot level south of the shaft and on the 1000-foot level north of the shaft where an ore shoot 80 feet long and 6 to 15 feet thick was mined (Logan, 1938, p. 233).

Placerville Gold Mining Company. Location: portions of secs. 6, 7, 17, 18, and 20, T. 10 N., R. 11 E., M. D., and sec. 36, T. 11 N., R. 10 E., M. D., extending from 1 mile south of the South Fork of the American River south through the Placerville city limits to Webber Creek, a distance of 4 miles. Ownership: Placerville Gold Mining Company, Box 191, Placerville, L. F. S. Holland, manager.

This concern owns a number of gold mines, both lode and placer, on the Mother Lode gold belt in and around Placerville. During the early days of the gold rush, these mines were the source of large amounts of gold. The present company was incorporated in 1911 and is a successor to the Placerville Gold Quartz Company, Ltd., an English concern that was incorporated originally in 1878. The best-known property of the group is the Pacific Quartz mine. Other notable mines of the group include the Epley Consolidated and True Consolidated.

The Pacific Quartz mine in the south portion of Placerville was operated from 1852-89, and again during 1914-15 (Logan, 1934, p. 35). Its total recorded production is $1,486,000 (Bowen and Crippen, 1948, p. 68). The Pacific vein is in a zone of talc and mariposite in Mariposa slate on the west side of the Mother Lode. Serpentine bodies are nearby. The vein strikes N. 25° W. and dips 70° NE. A number of ore shoots several hundred feet long and as wide as 12 feet were developed. The ore, which yielded $6 to $18 per ton in gold, contains appreciable amounts of arsenopyrite and silver. The mine is developed by a 700-foot shaft and 1,365-foot winze sunk from the 700-foot level. The ore was treated in a 20-stamp mill.

A consolidation of the Rose, Chester, Ida, Oregon, and Oregon Extension claims within the Placerville city limits was operated up until around 1888 (Irelan, 1888, p. 182). These claims were located at various times between 1865 and 1880. Gold-bearing quartz veins with arsenopyrite and galena in slate were developed by a 250-foot adit and a 225-foot inclined shaft. The ore was treated in a 10-stamp mill.

Just north of Placerville in Big Canyon is the True Consolidated mine, which includes the Young Harmon, Old Harmon, Halleck, and Berry claims. During the 1880's the property was operated by the True Consolidated Mining and Milling Company (Irelan, 1888, p. 180). These claims passed into the hands of the Placerville Gold Mining Company in 1893 (Logan, 1934, p. 26). The ore was treated in a 10-stamp mill. There are two northwest-trending veins in slate on the Mother Lode. The east vein has a greenstone hanging wall. The east vein averages 4 feet in width; the west averages 15 feet (Crawford, 1894, p. 125). Six ore shoots were known. The mine is developed by a 1,400-foot south drift adit. Also there is a 560-foot shaft on the

Old Harmon claim. In 1931, a small amount of prospecting was done on this claim (Logan, 1934, p. 26).

The Van Hooker, Grass, Brown Bear, Cinnamon Bear, White Bear, and Eureka claims lie north of the True Consolidated mine on the Mother Lode and are also owned by this company. These were worked prior to 1888 (Irelan, 1888, p. 181) and are extensively developed by underground workings. From 1928 to 1931, the Van Hooker mine was operated, and an ore shoot 80 feet long was mined. The ore yielded $7.25 to $27 per ton (Logan, 1934, p. 26). This mine is developed by a 1,200-foot drift adit and a winze. The ore was treated in a 10-stamp mill.

South of Placerville between Chili Ravine and Webber Creek is the Epley Consolidated mine, which was active up to 1888. This property consists of the Epley, Faraday, Henrietta, and Mammoth claims, which were located in 1867 (Irelan, 1888, p. 186). The vein in these properties strikes northwest, dips 65° NE., and averages 6 feet in width. Country rock is Mariposa slate. Two ore shoots, 125 and 150 feet long, were worked. The mine is developed by an adit and two shafts. The ore was treated by amalgamation, and a sulfide concentrate was sent to a chlorination plant.

Pyramid (Gold Reserve) Mine.   Location: secs. 12 and 13, T. 10 N., R. 9 E., M. D., 4 miles due north of Shingle Springs and 2 miles southeast of Deer Valley. Ownership: Rhoads Grimshaw, Auburn.

The Pyramid gold mine was active during the 1890's, and the ore was treated in a five-stamp mill (Crawford, 1896, p. 154). An ore shoot over 500 feet long was developed. At this time the mine was worked through two 500-foot drift adits and a 50-foot shaft. After lying idle for many years, the mine was reopened in 1933 (Logan, 1935, p. 23). It was operated for the following 6 years by several concerns. The value of the total output of the mine is approximately $1,000,000 (Rhoads Grimshaw, personal communication, 1956).

This deposit is in the West Gold Belt. Country rock is amphibolite schist, lying between gabbro-diorite on the west and a band of serpentine on the east. Ankerite and mariposite are present in the wall rock. The main quartz vein, which ranges from a few to as much as 20 feet in width, strikes northwest, and dips 40° to 60° NE. The gold is associated with pyrite and smaller amounts of galena, arsenopyrite, and chalcopyrite. The mine is developed by an 818-foot inclined shaft. The later production was chiefly from the 500-, 700-, and 800-foot levels.

From 1933 to 1937, the ore was treated in a cyanidation plant equipped with a mercuric generator in which the formation of mercuric cyanide hastened the cyanidation process (Logan, 1938, p. 245). After 1938, the ore was treated in a 10-stamp mill and concentrated by tabling.

Rosecranz (Rosecrans) Mine.   Location: sec. 21, T. 12 N., R. 10 E., M. D., 1½ miles northwest of Garden Valley. Ownership: Arthur S. Morey, 91 Rico Way, San Francisco.

This gold mine originally was worked prior to 1888 (Irelan, 1888, p. 171). At that time, it was worked to an inclined depth of 200 feet and the ore was treated in a 10-stamp mill. Around 1888, about $21,000 was produced from ore that yielded $10 per ton (Logan, 1938, p. 248). Some work was done on the property between 1916 and 1918. In 1936, the mine was reopened by the Lode Development Company of Auburn and was active until 1939.

Native gold and pyrite occur with quartz in a vein that lies in a 250-foot belt of amphibolite schist enclosed in Mariposa slate. The quartz occurs in irregular bodies. The vein strikes north, dips west, and averages $3\frac{1}{2}$ feet in width. Eighty percent of the gold is in the native state (Logan, 1938, p. 249).

The Lode Development Company deepened the old shaft to 350 feet. There are levels at 100, 130, 200, 250, and 350 feet. Open stopes were used. Ore shoots ranging from 125 to 150 feet in length were worked both north and south of the shaft. On the 100-foot level a 165-foot crosscut was extended east to the adjoining Taylor vein (Logan, 1938, p. 249).

In the last operations, the ore was treated in a 100-ton mill that was equipped with a Bendelari jig and a bank of five flotation cells. Thirty men worked at the mine and mill.

Shaw (Shan Tsz, Volo) Mine. Location: sec. 21, T. 10 N., R. 10 E., M. D., 2 miles north of El Dorado and 4 miles southwest of Placerville. Ownership: Evelyn R. Purser, 110 Bank Building, South San Francisco; leased by Volo Mining Company, 464 Main Street, Placerville, F. V. Phillips, president, Joseph Pickering, mill foreman.

This is one of the better-known gold mines of the West Belt in El Dorado County. For many years it was worked as an underground mine. However, recently all of the ore has been mined from an open pit. It was active during the 1880's (Irelan, 1888, p. 193) and again around 1915 (Tucker, 1919, p. 297). The Volo Mining Company leased the property originally in 1940 (F. V. Phillips, personal communication, 1953). During 1941 and 1942, some gold ore was mined and milled on an experimental basis. The mine was shut down in 1942, but copper ore from several properties in the county was treated at the mill, which is known as the Volo mill, in 1943-44. The mine was reopened in 1946 and operated almost continuously until 1953. The mill is being used at the present time to treat copper ore from the Copper Hill mine in Amador County (see Volo mill in Copper section).

Gold mineralization has taken place in a body of quartzitic schist that lies in slate and schist of the Calaveras group. The deposit often has been referred to as a quartzite dike or rhyolite dike, but according to Fairbanks, the Shaw deposit genetically occupies an intermediate position between the Big Canyon deposit to the south, a true replacement deposit, and the ordinary fissure quartz vein deposit (Fairbanks, 1894, p. 480). Pyrite, calcite, and talc are associated with the gold. The ore is low grade, containing $2 to $4 per ton (Joseph Pickering, personal communication, 1953). The deposit contains seams of secondary albite and quartz as well as a few thin basic dikes. Rich pockets of free gold have been found near the contact of the ore body and the surrounding country rock (Tucker, 1919, p. 297).

The ore body strikes N. 10° E., and dips 85° SE. It averages about 100 feet in width and has been worked for a distance of about 1,000 feet along the strike. Prior to 1915, the mine was worked by underground methods. It was developed by a 135-foot shaft, a 400-foot crosscut adit, a 300-foot south drift, and a 200-foot north drift. After 1940, the mine was worked in an open cut that now is about 1,000 feet long, 100 to 150 feet wide, and 20 to 50 feet deep. The ore was blasted from the pit faces, loaded with a power shovel, and trucked to the mill.

The ore was treated by amalgamation flotation, and cyanidation at the mill which has a capacity of 300 tons. Five men worked at the mine and mill.

Skipper (Esperanza) Mine. Location: sec. 7, T. 12 N., R. 10 E., half a mile east of Greenwood. Ownership: H. R. Swartz, Greenwood.

This unpatented mining claim, which was once known as the Esperanza mine, has been worked intermittently by Mr. Swartz since 1935. The deposit consists of gold-bearing meta-andesite which is highly weathered and sheared. There is also some fresh greenstone containing auriferous pyrite. The gold is finely disseminated through the greenstone.

The zone of gold mineralization is 60 feet long and 40 feet wide and has a nearly east trend, although the country rock trends north. It is developed by an open cut several hundred feet long and a 50-foot, north-dipping inclined shaft sunk in the open cut. There are numerous prospect holes and cuts on the property.

Sliger Mine. Location: sec. 36, T. 13 N., R. 9 E., M. D., 1 mile west of Spanish Dry Diggings. Ownership: Sliger Gold Mining Company, 3628 Fulton Street, San Francisco; H. W. Simpers, president.

The Sliger gold mine was worked originally in 1864, and during the 1870's the ore was treated in a five-stamp mill on the property (Logan, 1926, p. 418). During these early operations the mine was worked to a depth of about 300 feet, and the production totaled $225,-000 (Logan, 1934, p. 38). The property was idle until 1922 when the Sliger Gold Mining Company was organized. The shaft was deepened, a 15-stamp mill was erected, and milling of ore began in 1929. In 1932, a new mill with a flotation unit was erected. The mine was operated until 1934 by the Sliger Gold Mining Company and from 1934 to 1937 by the Middle Fork Gold Mining Company. In 1937, the Mountain Copper Company leased the mine and did some exploration work (Logan, 1938, p. 250). From 1938 to 1942, the mine again was operated by the Middle Fork Mining Company. The mine has been idle since 1942. In 1953 most of the surface plant was sold. From 1932 to 1942, 309,000 tons of ore were mined from which $2,625,000 was recovered (H. W. Simpers, personal communication, 1955).

The deposit is in a fault zone between Mariposa slate on the footwall and a band of serpentine on the hanging wall. Country rock is amphibolite. The ore is composed of siliceous slate containing numerous quartz veinlets with free gold, fine disseminated pyrite, and smaller amounts of arsenopyrite. Ankerite is present. The ore zone, which averages 30 feet in width, strikes north and dips 63° E. The ore shoots are lensoid in shape. However, they do not have definite walls but merge into lower-grade rock. It is believed that the gold-bearing solutions rose along several favorable bands in the slate to the fault zone, where they spread out to form the ore shoots (Logan, 1938, p. 250).

About 40 percent of the gold is free, as high as 925 in fineness, and about 60 percent is contained in sulfides (H. W. Simpers, personal communication, 1955). The tenor of the ore remained virtually unchanged from the 300-foot level down to the 2,000-foot level. The mine is developed by a 2,000-foot inclined shaft. Ore was mined in open stopes that later were filled with mill tailings.

At the mill the ore was treated by two-stage crushing and then sent to ball mills. From the mills, the pulp went to a Diester table, and table tailings were treated by flotation in six rougher and then two cleaner cells. During the last years of operation, an average of 150 tons of ore per day were mined (H. W. Simpers, personal communication, 1955).

Stuckslager Mine. Location: sec. 24, T. 11 N., R. 9 E., M. D., 1 mile southwest of Lotus. Ownership: Beryl E. McKenney, 31 Lake Vista Avenue, Daly City 25.

This gold mine was worked originally by the Stuckslager family during the 1860's. The property was purchased during the 1880's by the McKenney family who have operated it intermittently ever since (Mrs. E. J. McKenney, personal communication, 1955). From 1935 to 1942, the mine was worked by D. J. McKenney of Sacramento. At the present time E. J. McKenney Jr. is intermittently sinking a shaft on the hill above the adit portal.

This deposit is noted for the occurrence of roscoelite, a vanadium mica. Roscoelite is a very rare mineral found in few other places in the United States (Bowen and Crippen, 1948, p. 68). Native gold occurs in pockets with greenish or clove-brown scales of roscoelite in a quartz vein. The main vein strikes northwest, dips southwest, and ranges from a seam to 2 or more feet in width. Also present are a number of thin quartz stringers. Country rock is greenstone and slate lying between serpentine on the west and granodiorite on the east. The mine is developed by a 500-foot drift adit and several shallow shafts. During the 1930's, the ore was treated in a two-stamp mill. The shaft now being sunk is about 50 feet deep.

Sugar Loaf Mine. Location: secs. 1 and 12, T. 8 N., R. 9 E., M. D., 2½ miles northeast of Latrobe. Ownership: Sugar Loaf Mining Company, Route 1, Box 57, Shingle Springs.

This is a pocket gold mine of the West Belt. It first was worked by open cuts during the early days of the gold rush. It was active again during the 1880's. The mine was prospected during the early 1920's. During the middle 1930's a 175-foot shaft was sunk, and some high-grade ore was mined (J. H. Wren, personal communication, 1955). Since 1954, the mine has been operated intermittently by the Butler Mining and Development Company of Sacramento, and some high-grade ore has been shipped to the smelter.

Native gold associated with varying amounts of pyrite and galena occur in a quartz vein averaging 5 feet in width. The vein strikes north and dips 75° W. The footwall consists of slate; the hanging wall is talcose serpentine. Associated with the vein on the footwall side are several dikes of medium-grained diorite. The mine is developed by a 175-foot inclined shaft with levels at 100, 130, and 175 feet. High-grade ore was found recently in the vein just below the 100-foot level. Three men work intermittently at the mine.

Taylor (Idlewild) Mine. Location: secs. 21 and 30, T. 12 N., R. 10 E., M. D., on the Mother Lode 2 miles northwest of Garden Valley. Ownership: W. N. Guth, 2024 North New England Avenue, Chicago, Illinois.

The Taylor gold mine originally was worked in 1865 (Logan, 1934, p. 41). It was active again from the late 1880's to about 1902 (Logan, 1923a, p. 210). Some work was done on the property during the period of 1939-41 by the Lode Development Company of Auburn. The estimated total output of the mine is valued at about $1,000,000 (Logan, 1934, p. 42).

A vein about 2 feet wide at the surface, but increasing in width to an average of 14 feet at depth, lies between a Mariposa slate hanging wall and a hard greenstone footwall. The vein strikes northwest and dips 50° NE. There is a considerable amount of gouge in the hanging wall. The gold, which was finely divided, was in ribbon rock, the best ore being near the hanging wall. The grade of the ore ranged from $4 to $8 per ton (Logan, 1934, p. 42). The mine is developed by a 1,225-foot inclined shaft with levels every 100 feet. Because of heavy ground, stopes were filled with waste rock. The ore was treated in a 40-stamp mill, and the concentrates were cyanided.

Union (Springfield) Mine. Location: sec. 12, T. 9 N., R. 10 E., M. D., 2 miles southeast of El Dorado on the Mother Lode. Ownership: El Dorado Mining Company, P. O. Box 1724, Spokane 7, Washington.

The Union mine is considered to have been the largest source of lode gold in El Dorado County. However, because very few production figures have been made available (Logan, 1934, p. 42), the total output of the mine is unknown. The estimated value of the total output has ranged from $2,700,000 (Bowen and Crippen, 1948, p. 66) to as much as $5,000,000 (Tucker, 1919, p. 299).

In the early 1850's, rich surface outcrops of the area were worked, and soon after a shaft was sunk. The Union mine and the Church mine, later worked separately, were consolidated and produced more than $600,000 prior to 1868 (Logan, 1934, p. 42). The Union mine was worked again from 1871 to 1886, and the ore was treated in a 15-stamp mill. From 1896 to 1909, it was operated by the Union Gold Mining Company, which enlarged the mill to 20 stamps. Some prospecting was done on the property during 1914-15. In 1934, the mine was reopened by the Gold Fields American Development Company (Logan, 1938, p. 252). The shaft was deepened to 2,000 feet and the lower workings were rehabilitated. During 1936-37, the mine was operated by the Montezuma-Apex Mining Company and the ore was trucked to the Montezuma Apex mill at Nashville. Between 1937 and 1940, mining operation was sporadic and inconsequential. The mine has been idle since 1940.

A number of veins that strike north-northeast and dip 60° to 79° E. are in Mariposa slate. The principal veins are the Poundstone (or East Gouge), which had the greatest production, the McCosmic vein about 200 feet to the west, and the Klondyke vein, which was developed north of the main shaft. These veins range from 5 to 10 feet in width. A series of lesser veins lie to the west of the main veins. Ore from the

Poundstone vein yielded $8 per ton; the ore from the McCosmic vein yielded up to $25 per ton (Logan, 1938, p. 253). Ore recovered from shallow workings during the early days often was considerably richer.

The mine is developed by the 2,000-foot vertical main (Springfield) shaft which cuts the Poundstone vein at 1,200 feet and the McCosmic vein at about 1,540 feet. About 750 feet north is the 900-foot Clement shaft and 200 feet beyond is the 500-foot Klondyke shaft. Also there are several crosscut adits driven to the west, one 700 feet long near the Springfield shaft and another to the north near the Klondyke shaft about 600 feet long.

Vandalia Mine. Location: sec. 19, T. 9 N., R. 10 E., M. D., 4 miles south of Shingle Springs on the west side of Big Canyon. Ownership: G. L. Tripp, Route 1, Box 5, Shingle Springs.

This gold mine was worked originally in 1885 and was again active in 1888. The ore was treated in a five-stamp mill (Irelan, 1888, p. 172). It was idle during the 1890's, but around 1900, a cyanidation plant, which treated accumulated tailings as well as ore from the mine itself, was erected (Storms, 1900, p. 97). A small amount of work was done on the property in 1926 and 1928. During 1936-37, the Page Consolidated Mining Company prospected the mine and erected a 150-ton mill and cyanidation plant. However, little gold was produced (Logan, 1938, p. 254).

The deposit consists of bodies of fine-grained silicified schist containing disseminated auriferous pyrite. Country rock is amphibolite schist and quartz porphyry. The oxidized zone extended to a depth of about 100 feet. The ore bodies were over 80 feet wide, as much as 300 feet long, and commonly cut the chistosity of the enclosing rocks (Storms, 1900, p. 97). The mine was developed by several drift adits and open cuts.

Veerkamp (Gold Coin) Mine. Location: sec. 33, T. 12 N., R. 10 E., M. D., 1 mile west of Garden Valley on the Mother Lode. Ownership: Ray A. Veerkamp, Garden Valley.

A small amount of shallow prospecting was done on this gold property in the early days (Logan, 1934, p. 37). In 1933, some ore was mined from an open cut that was treated at the Beebe mill, and an adit was driven. From 1935 to around 1940, the property was operated by the Gold Company, Ltd., a Canadian concern. There was a small amount of work done in the mine again in 1950.

Several quartz veins as well as a number of quartz veinlets occur in highly weathered schist and slate. A large low-grade vein strikes N. 25° W. and intersects a smaller but higher-grade vein that strikes N. 15° E. A diabase dike lies just to the east. The ore contained about $12 per ton; some of the country rock enclosing the quartz veinlets contained up to $4 per ton (Logan, 1938, p. 247). Several pockets of high-grade ore were mined during the 1930's.

The mine is developed by a 180-foot shaft with levels at 60 and 97 feet. There are 1,200 feet of drifts on the 60-foot level and about 1,000 feet on the 97-foot level. Also there are several adits and a number of open cuts. Various methods of treating the ore were tried in a mill on the property, including flotation and cyanidation.

Zantgraf (Montauk Consolidated, Zentgraf) Mine. Location: sec.
5, T. 11 N., R. 8 E., M. D., 1 mile south of Rattlesnake Bridge on the
ast side of the American River, 6 miles southwest of Pilot Hill. Owner-
iip: United States of America (land withdrawn for dam site).

The Zantgraf gold mine first was worked in 1880 (Logan, 1938, p.
55) and in 1888 a 10-stamp mill was in operation (Irelan, 1888, p.
01). It was active during the 1890's, and the mill was enlarged to 20
amps (Crawford, 1894, p. 162). Between 1898 and 1901 it was oper-
ted by the Montauk Consolidated Mining Company. The total output
or the mine up to 1901 was valued at about $1,000,000 (Logan, 1938,
. 256). The property was idle until 1924 when a small amount of pros-
ecting was done just north of the mine. From 1933 to 1938 and again
1 1941, W. B. Longan prospected the property and some gold was
roduced. Several shallow shafts were sunk and a considerable amount
f drifting was done. The mine has been idle since.

There are several nearly parallel veins in granodiorite and amphib-
lite. Amphibolite lies to the east. The veins strike northwest, dip
8° to 50° SW., and are accompanied by diorite dikes. The mine work-
igs are largely in granodiorite. Nearly all ore that was mined was
rom the main or Zantgraf vein, which ranges from 2 to 6 feet in width.
'he ore contained free gold with associated auriferous pyrite and
alena. The values of the ore ranged from $2 to as much as $100 per
on, but averaged between $7 and $8 per ton (Logan, 1938, p. 256).

The mine is developed by a 1,130-foot inclined shaft with levels at
very 100 feet. There is a 500-foot crosscut adit that joins the shaft
n the 300-foot level. An ore shoot was stoped from the 300-foot level to
he surface for a distance of 900 feet. Another ore shoot 600 feet long
as developed in the lower levels. There is also a 200-foot shaft 900 feet
orthwest of the main shaft with levels at 80 and 180 feet, and a 190-
oot shaft on the Montauk vein, both of which were sunk in the 1930's.

The ore was originally treated in a stamp mill. During the 1930's a
0-ton Chilean mill was used; it was later replaced by a ball mill.

Placer Deposits

Placer gold deposits in El Dorado County have been worked by
lredging, hydraulicking, and drift mining. Most dredging has been
lone with dragline dredges, although a few connected-bucket dredges
vere used at one time. Hydraulic mining consists of directing large jets
f water under high pressure toward banks of gold-bearing gravels. The
xcavated gravels flow through strings of sluice boxes where the gold
s recovered. Buried stream channels of Tertiary age have been de-
eloped by underground mines known as drift mines. Drifts and cross-
uts are run along pay streaks in the old river channels. The excavated
ravel is brought to the surface and sluiced or sent through a washing
lant to recover the gold.

Placer gold deposits in El Dorado County range from Eocene to
Recent in age. The gold in all of the placer deposits originated from
veins in the granitic and metamorphic rocks of the Sierra Nevada.
When those veins were subjected to prolonged erosion and weathering,
he gold was released and became intermingled with other weather-
resistant material in streams. It was then concentrated by the action
f running water usually at or near the base of the stream deposits.

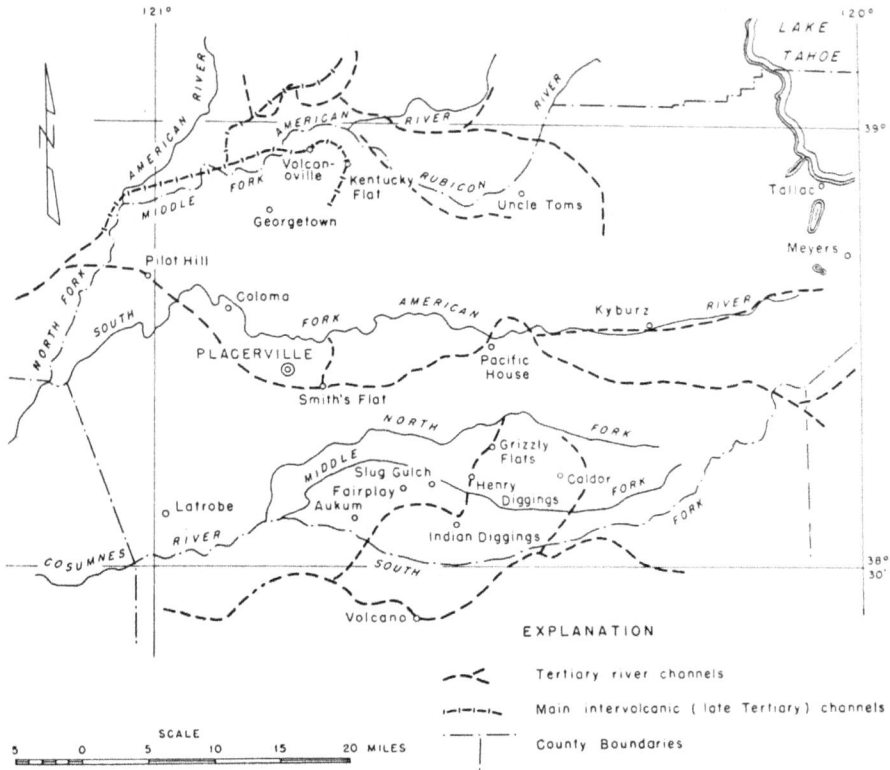

FIGURE 21.   Map of El Dorado County showing Tertiary river channels
(after Lindgren 1911, and Jenkins 1932).

Associated with the gold in placer deposits are magnetite and other
heavy resistant minerals such as ilmenite, garnet, rutile, zircon, and
platinum-group metals. The most productive placer deposits in the
county were in Tertiary channels in the Placerville area.

The present river and stream systems that developed in the Quaternary period swept away much of the Tertiary volcanic cover and
cut new canyons in the bedrock. Some of the canyons are several
thousand feet below the old surface. This tremendous amount of erosion
and natural concentration of heavy materials resulted in accumulation
of large quantities of gold in the present streams and their branches.
These deposits range in size from those that are small but often rich in
tributary creeks to the large dredging fields many square miles in
extent that lie west of El Dorado County. Small rich surface deposits
in tributary creeks were responsible for the very large amounts of gold
produced during the first few years of the gold rush. During the 1930's
there was considerable placer mining in El Dorado County along the
American and Cosumnes Rivers and many of the creeks. At most of
these operations, draglines and floating washing plants known as
"doodlebugs" were used.

In Eocene times, the American and Mokelumne Rivers occupied different channels than they do now. Gold was concentrated in these
ancient ancestral channels, some of it in rich deposits. Volcanism that

began at or near the end of the Eocene epoch and continued into the Pliocene epoch buried these channels under great thicknesses of volcanic material. Several times during the period of volcanism, the rivers were forced into new channels, which now are known as "intervolcanic channels." The intervolcanic channels generally are leaner in gold than the early Tertiary channels. Portions of the ancient channels were preserved by the thick cover of volcanic rocks and are exposed on the sides of the present stream canyons.

Detailed descriptions of the Tertiary channel deposits as well as a discussion of their geology are contained in U. S. Geological Survey Professional Paper 73 (Lindgren, 1911). Information also can be found in *Auriferous gravels of the Sierra Nevada* (Whitney, 1880), California Mining Bureau Bulletin 92 (Haley, 1923), Geologic map of the northern Sierra Nevada (Jenkins, 1932), *California's gold-bearing Tertiary channels* (Jenkins and Wright, 1934), and California Division of Mines Bulletin 135 (Averill, 1946).

The most productive Tertiary channel deposits in El Dorado County were in the Placerville area, east and northeast of Georgetown, and in the Grizzly Flat-Fairplay-Indian Diggings region. These areas are described separately in this report, and the individual mines are described in the tabulated list under *Placer gold*.

Georgetown Area. An intervolcanic channel of the Middle Fork of the American River flowed north from Tipton Hill, 5 miles northeast of Georgetown, to the old mining camp of Kentucky Flat. At Kentucky Flat the main or "white" channel, which is 350 to 500 feet wide, is known for the abundance of large quartz boulders and the coarseness of the gold. Several other intervolcanic channels exist locally that differ in direction from the main channel (Lindgren, 1911, p. 168). Probably the best-known operation in the Kentucky Flat area was that of the Two Channel Mining Company, which operated extensive drift and hydraulic mines for several miles along the channel during the 1890's (Crawford, 1896, p. 160). The principal sources of gold were the Kenna, Kentucky Flat, Morris, and Tiedeman mines. Gravel at the Kenna mine was well cemented and was treated in a 10-stamp mill.

From Kentucky Flat, the channel turned west and flowed through Volcanoville. Considerable amounts of gold were produced from drift and hydraulic mines in the Volcanoville area. In places, such as at the Buckeye Hill mine on Buckeye Point, the channel was as much as 1000 feet wide (Crawford, 1894, p. 105). This mine was active during the 1890's and again during the 1930's. Other notable sources of gold in the area included the Channel Bend, Gray Eagle Cliff, and Rubicon mines. Gold also was mined from isolated patches of Tertiary gravel capping several peaks southwest of Volcanoville and north of Georgetown. These include Bottle Hill, Cement Hill, and Jones Hill. This channel continued on into Placer County southwest of Foresthill. Two other branches of the Tertiary Middle Fork lie to the east : one extends west from near Uncle Tom's and northwest along Peavine Point east of Kentucky Flat; the other extends in a north-northwest direction east of Uncle Tom's. Little gold has been mined from these two channels.

Mokelumne River Deposits. A channel of the Tertiary Mokelumne River flowed south from just south of Baltic Peak to Grizzly Flat. At

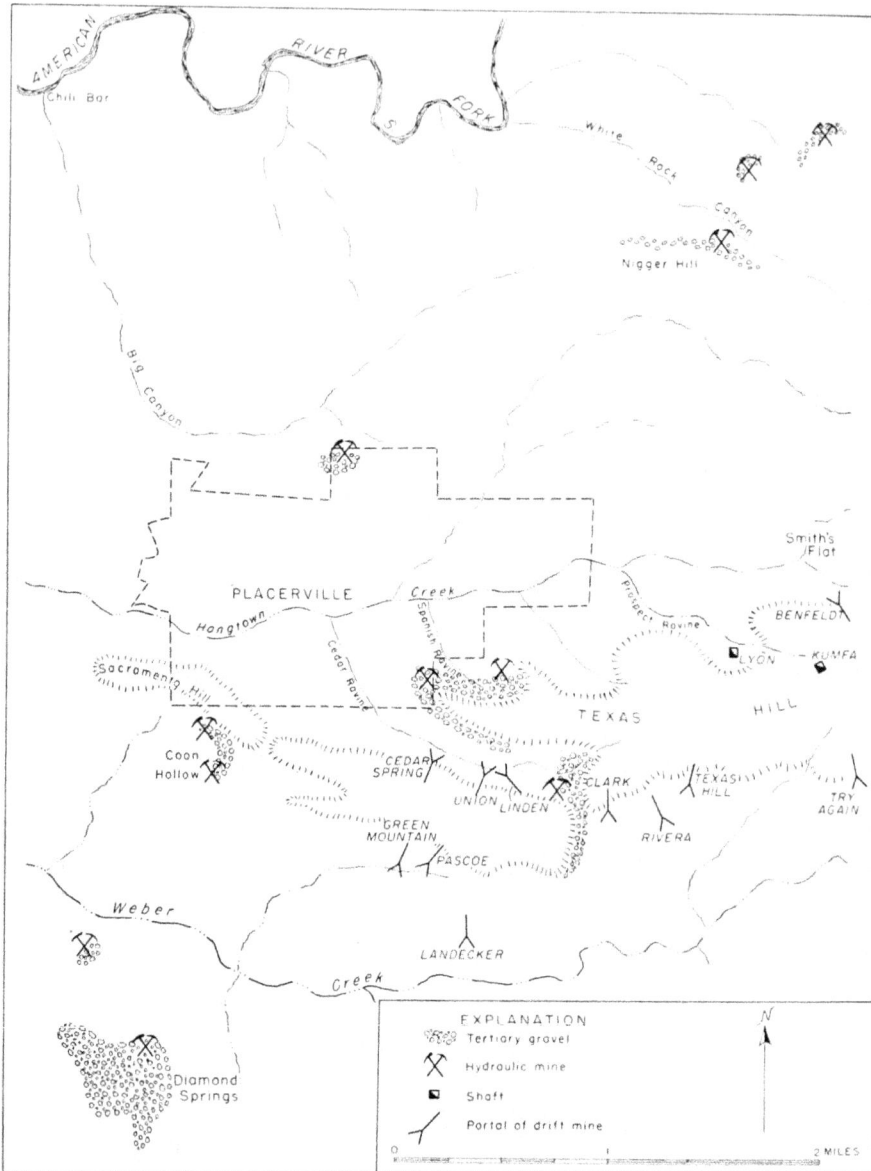

FIGURE 22.   Sketch map of Placerville area, showing locations of former placer mines and Tertiary gravel deposits.

Grizzly Flat, gold-bearing gravels underlie tuffaceous andesite and rest on granitic bedrock. The gravels are thin and the channel narrow (Lindgren, 1911, p. 180), but substantial quantities of gold were produced from drift and hydraulic mines in this district in the past. The Grizzly Flat mine, which was active during the 1880's, 1890's, and again in 1918, was an important source of gold.

This channel appears again at the old mining camp of Henry Diggings, a few miles to the southwest. The gravels here were rich (Lindgren, 1911, p. 180). Notable sources of gold in this area were the Armstrong, Carrie Hale, Irish Slide, Last Chance, and Payne drift mines. Continuing south, the main channel flowed just west of Brownsville, another old mining camp one mile south of Omo Ranch, and then southwest to a point about a mile north of Indian Diggings.

Although Indian Diggings is not on the main channel, gravel deposits here were extremely productive. Bedrock is a wide belt of limestone which contained numerous rich potholes. The Hayward mine, which is a consolidation of several drift and hydraulic mines, was a major source of gold in the district. As late as the early 1900's, the annual gold output for the district was $7,000 to $10,000 (Lindgren, 1911, p. 181). During the 1920's the Jinkerson drift mine was active (Logan, 1926, p. 439), and the Ball drift mine was worked in the 1930's (Logan, 1938, p. 217).

Auriferous gravels are present in the Fairplay area, although they too are not part of the main channel. A channel that extended from Slug Gulch southeast to Fairplay was mined by hydraulicking for many years. The California Mohawk Mining Company operated properties that extended several miles along this channel (Tucker, 1919, p. 301). Recently some radioactive material was discovered at Slug Gulch (see section on Radioactive Minerals).

There was another south-flowing channel of the Tertiary Mokelumne River east of Caldor. Some gold was mined from isolated gravel deposits along this channel by hydraulicking (Lindgren, 1911, p. 183).

Placerville Area. The most productive Tertiary channel deposits in the county were in the Placerville area. This area includes the deposits at Coon Hollow, Diamond Springs, Smith's Flat, Texas Hill, and White Rock. The value of the total gold output for the district has been estimated to be about $25,000,000 (Haley, 1923, p. 144). The Tertiary South Fork of the American River entered the area from Newtown. There are numerous tributaries in the area, some of which have yielded large amounts of gold. Nearly all are overlain by rhyolite tuff; bedrock is Mariposa slate, Calaveras metasediments, and granodiorite.

The best known tributary is the Deep Blue Lead which has been traced south from White Rock Canyon to Smith's Flat and southwest to Cedar Ravine. This tributary follows approximately a slate-granodiorite contact for several miles and is believed to have derived much of its gold from many small veins in the contact zone. It was mined extensively between the late 1860's and the 1890's. The gravels at White Rock yielded $5,000,000 while the Lyons mine just south of Smith's Flat yielded $1,400,000 (Lindgren, 1911, p. 172). The latest work of any extent in the Smith's Flat area was at the Hook-and-Ladder mine which was operated almost continuously from 1918 to 1932 (Averill, 1946, p. 255).

Rich gravels were mined in the area between Coon Hollow and Texas Hill. Drifting was done at Coon Hollow from 1852 to 1861, and the hill was hydraulicked from 1861 to 1871. This area was the source of about $10,000,000 (Lindgren, 1911, p. 172). The Excelsior claim in Coon Hollow yielded $5,000,000 from an area 20 acres in extent (Whitney, 1880, p. 119). The gravels contained as much as $5 per yard. At present accumulated tailings in Coon Hollow are being mined for sand and gravel by the Harms Construction Company (see Stone section).

The Green Mountain and Cedar Spring channels lying east of Coon Hollow were extensively mined by drifting. Gravels at Spanish Hill were the source of gold valued at $6,000,000 (Lindgren, 1911, p. 172). The Deep Blue Lead, which continued in a southwest direction from Smith's Flat was developed by the Linden and Landecker drift mines. Other productive mines in this portion of the area included the Clark, Ditch Company, Rivera, Texas Hill, and Try Again mines.

Gold Dredging. There have been a large number of gold dredging operations in El Dorado County. The majority of these were the dragline dredges or "doodlebugs" which were active during the 1930's and 1940's. At present, there is little gold dredging in the county.

Among the earlier (1914-18) dredging operations in El Dorado County were those of El Dorado and Placer Counties Gold Mining and Power Company, which operated a floating dredge with 3½-cubic-foot buckets on the Middle Fork of the American River near Poverty Bar, and the Pacific Gold Dredging Company, which operated a floating dredge with 7½-cubic-foot buckets at Mammoth Bar (Logan, 1918, p. 29). A dragline dredge was operated at the Jumbo Placer mine on Carson Creek 1 mile west of Clarksville in 1923 (Logan, 1923b, p. 142). In this operation, gravels were excavated by a dragline and delivered to a hopper which fed a sluice box.

During the period of 1924-25, the Boles placer claim on the American River upstream from Rattlesnake Bridge was prospected by a combined diving and pumping operation (Logan, 1926, p. 428). A diver at the bottom of the river directed the nozzle from a barge-mounted suction pump that discharged into sluice boxes. Large boulders were removed by a scoop lowered from an overhead cable spanning the river.

Between 1933 and 1941, dragline dredging methods and equipment were greatly improved. This, coupled with a rise in the price of gold in 1934, caused a great increase in the production of placer gold in El Dorado County during this period. Also, a lower capital outlay and portability of a dragline dredge made it possible to work many deposits that were too small to justify the construction of a connected-bucket dredge. Many of the creeks and gravel deposits in western and central El Dorado County were dredged by these outfits during the 1930's.

Dragline dredges or "doodlebugs" consist essentially of two parts, a dragline excavator and a washing plant. Draglines used in this type of work are equipped with buckets ranging from 1 to 3 cubic yards in capacity. Floating washing plants are mounted on a barge and consist of a hopper into which the gravel is delivered, a trommel, a belt conveyor to stack the coarse tailing, and banks of riffles. Usually a tractor with a scraper is used to clear the way ahead for the dragline. Draglines are most successful where the gravel is not much more than 20 feet deep and not too compact or well cemented. Published production figures for El Dorado County show recoveries have ranged from 15 to 30 cents per cubic yard.

A few non-floating washing plants have been employed. These washing plants are mounted on tractor treads and usually are self-propelled. Also, a few suction dredges have been employed. These consist of a barge-mounted pump with pipe, a digging nozzle, and a washing plant mounted on the barge.

According to U. S. Bureau of Mines records, at least 33 dredges have been active in the county since 1938. Most of these have been dragline dredges with floating washing plants. All known dredging operations are in the tabulated list in the back of this report (see placer gold in tabulated list).

The following are some of the better-known gold-dredging operations of recent years in El Dorado County:

Big Canyon Dredging Company of Fresno operated a dragline with a 3-cubic-yard bucket in the Deer Creek area, near Shingle Springs, from 1938 to 1943. As much as half a million yards of gravel per year was handled (Averill, 1946, p. 255); recoveries averaged about 16 cents per yard.

El Dredging Corporation of Greenwood operated a dragline dredge with a 1½-cubic-yard bucket in the Coloma and Greenwood areas in 1940-41, near Georgetown in 1946-47, and near Camino in 1948.

General Dredging Corporation of Natoma operated a 1½-cubic-yard dragline dredge and a second 2-cubic-yard dragline dredge in the Coloma area from 1939 to 1942 and near Shingle Springs in 1941.

Greenhorn Dredging Company of Auburn operated a 2-cubic-yard dragline dredge on the Middle Fork of the Cosumnes River near Youngs during 1940-42 and again in 1947.

Lord and Bishop of Sacramento operated a 3-cubic-yard dragline dredge and Bodinson floating washing plant on Greenwood and Carson Creeks in 1949 and 1950.

The River Pine Mining Company, Ltd., of San Francisco operated a 2-cubic-yard dragline and floating washing plant on the Middle and North Forks of the Cosumnes River during 1941-42, 1946, and 1949-50.

Seam Deposits

El Dorado County's "seam diggings" or seam gold deposits are in the north portion of the Mother Lode belt north of Placerville. They are found in Mariposa slate as well as in amphibolite and chlorite schist. The gold is in narrow quartz veinlets in the country rock in definite zones. These zones, which are as much as several hundred feet wide, contain quartz veinlets in several systems which have definite dips and strikes. The richest portions of seam deposits are for the most part near the intersections of two or more systems of veinlets, or where one system intersects a larger quartz vein (Logan, 1934, p. 43).

The surface portions of the seam deposits have been subjected to long-continued weathering. As rock disintegration by mechanical breakdown and chemical weathering progressed, the softer and more soluble portions of the rock were carried away by erosion, leaving the quartz veinlets and gold. As this process continued, the surface portions of these deposits became enriched. The name "residual placer" has been applied to this type of deposit (Jenkins, 1935, p. 158).

The soft, weathered, upper portions of seam deposits first were mined by hydraulicking and sluicing. After the residual material had been removed, the hard, unweathered, mineralized portions of the deposits were encountered and mined as lode-gold deposits by conventional underground methods.

Seam deposits were worked extensively by hydraulicking during the 1860's and 1870's. Old records show the surface portions contained about $1 per cubic yard in gold (Logan, 1934, p. 44). The largest and most productive seam deposits were in the Georgetown, Spanish Dry Diggings, Greenwood, and Garden Valley areas. The seam mines at Georgia Slide northwest of Georgetown had the highest total gold output of the seam deposits.

Georgia Slide (Beattie, Blue Rock, Mulvey Point, Pacific, Parsons) Mine. Location: sec. 3, T. 12 N., R. 10 E., and secs. 34 and 35, T. 13 N., R. 10 E., M. D., 1½ miles north-northwest of Georgetown on the south side of Canyon Creek. Ownership: Kelsey Lumber Company, 515 Main Street, Placerville.

The best-known and largest seam gold deposit in the county, the Georgia Slide mine, includes the Blue Rock, Beattie, Mulvey Point, Pacific, and Parsons claims. These claims originally were worked separately. The Beattie claim was mined by hydraulicking continuously from 1853 to 1895 and the Blue Rock and Pacific claims continuously from 1856 to 1895 (Logan, 1934, p. 46). After the cessation of hydraulic mining, small amounts of ore were treated in several stamp mills that had been erected on the property. In 1915, a 10-stamp mill was being operated to determine the gold content of the large amount of accumulated tailings (Knopf, 1929, p. 48). Later a number of lessees worked the quartz stringers in search of high-grade pockets.

As the value of the estimated total output of the seam mines is $3,000,000 (Knopf, 1929, p. 49), and about $3,500,000 was believed to have been taken from placer deposits in Oregon Canyon and Canyon Creek, which derived their gold from the seam deposits (Logan, 1934, p. 46), this deposit has been the source of at least $6,500,000 in gold. It is estimated that about 3,500,000 cubic yards were removed from the immense open pit that now remains (Knopf, 1929, p. 49).

The deposit is a lenticular body of highly weathered green chloritic siliceous schist with some interbedded black slate. It strikes N. 5° W. and dips 80° E. The seams are gold- and pyrite-bearing quartz stringers ranging from less than an inch to a foot or more in thickness. The average stringer was 1 to 2 inches thick. Some of the gold mined from this deposit was beautifully crystallized. Some stringers swelled into lenses several feet thick. In some places albite and calcite are present in abundance. According to Preston, the material that was mined contained $2 to $3 per ton in gold (Preston, 1892, p. 203). There were five belts of schist in which the stringers were more abundant than elsewhere (Knopf, 1929, p. 40). The easternmost belt was 35 feet thick.

Originally the deposit was mined by hydraulicking. Later the banks were excavated by blasting, and the debris was hydraulicked into sluices. Whenever the quartz seams were encountered they were treated by hand mortaring (Crawford, 1894, p. 102). At present, there is a huge open pit about 1000 feet long and nearly 600 feet wide at the top. Water for hydraulicking was obtained from Canyon Creek, North Canyon, and Dark Canyon.

Hart Mine. Location: sec. 27, T. 12 N., R. 10 E., M. D., between Manhattan and Empire Creeks 1 mile north of Garden Valley. Ownership: Daniel Gellermann, 4524 Brand Way, Sacramento.

The Hart gold mine is in the seam belt just north of the split in the Mother Lode belt at Garden Valley. In the early days, the property was worked by hydraulic mining (Logan, 1934, p. 47). Later the deposit was prospected to determine the feasibility of erecting a mill. In 1930, a mill was erected (Logan, 1938, p. 233), and the property was worked intermittently until 1939.

This deposit is in the same belt of amphibolite schist that contains the Alpine and Black Oak gold mines. Four parallel systems of veins and stringers have been developed. The average gold recovery of ore mined in this property for the 4 years prior to 1938 was $5 per ton (Logan, 1938, p. 233).

The mine is developed by an open pit 175 feet long, 50 feet wide, and 40 feet deep, two drift adits, and two shafts, 200 and 300 feet in depth. The ore was treated in an eight-stamp mill.

Iron

Iron ore is found in several localities in El Dorado County, but there has been no recorded production. At the Reliance deposit north of Bass Lake in the western portion of the county, there are two vein-like bodies of magnetite several feet thick in coarse-grained gabbro. At the Chaix prospect near Latrobe, there is a lens of magnetite and hematite with a considerable amount of silica.

Lead

Small amounts of lead have been produced nearly every year as a by-product of gold and copper mining in the county. The primary mineral galena (PbS) is the principal ore mineral. In El Dorado County, galena occasionally is a minor constituent of Mother Lode gold ores and foothill copper ores. In a few cases, galena is a major constituent of East Belt gold ores. At present most of the lead produced in the county is from the Hazel Creek gold mine, where considerable amounts of galena are present in the ore. The galena is recovered by selective flotation.

Manganese

Although there has been no recorded output of manganese ore in the county, several small deposits were prospected during World War I and again during World War II. The largest of these were the Alderson deposit $1\frac{1}{2}$ miles south of Placerville, where manganese oxide in sandstone interbedded with Tertiary volcanic rocks crops out for a distance of 150 feet, and the David deposit west of Georgetown, where thin veins of manganese oxide are found in chert and quartzite. Other prospects include the Martinez and Mocettini deposits (see manganese section in tabulated list). Most of the manganese of the Sierra Nevada was originally deposited with silica as a silicate or as a carbonate. Manganese silicates (bementite and rhodenite) and carbonates (rhodochrosite) are the primary minerals which subsequently weather to black manganese oxides (pyrolusite and psilomelane).

Molybdenum

Small amounts of molybdenum minerals have been found in El Dorado County, but there has been no recorded production. At the Cosumnes copper mine, molybdenite (MoS_2) and powellite ($CaMoO_4$) occur in the tactite associated with bornite and chalcopyrite.

Nickel

Garnierite, an ore mineral of nickel in several places in the world, has been found in lateritic soils at the Pillikin chrome mine. Also it has been reported near Lotus (Murdoch and Webb, 1948, p. 155). However, nickel has not been produced in the county.

Garnierite, an amorphous hydrous magnesium nickel silicate, is found in rounded masses with smooth surfaces or as earthy fillings in veins or in cavities in zones containing silica veinlets in decomposed bedrock. Such deposits originated by residual concentration resulting from decay of ultrabasic rocks such as serpentine, under intense weathering conditions. Almost all ultrabasic rocks contain some nickel. However, only under proper conditions of weathering are these rocks decomposed and the nickel and some of the silica concentrated in the soil mantle. These soils are known as laterites and occur in the serpentine areas of El Dorado County. These are the most favorable areas in the county for nickel.

Platinum-Group Metals

Minor amounts of platinum and related metals have been recovered sporadically in El Dorado County as a by-product of placer-gold operations. Platinum-group metals are found as alloys in smooth thin flakes and small rounded nuggets in placer deposits. The proportion of these metals to gold is usually small. Platinum is believed to have been originally deposited by magmatic processes in basic and ultrabasic rocks. In placer mining it usually is recovered by jigging and panning the black sand concentrate after the gold has been removed.

Quicksilver

Some quicksilver was produced in El Dorado County in the 1860's at the Bernard cinnabar mine two miles west of Nashville (see tabulated list). The mine was prospected in 1903 and again in 1917, although there is no later record of production (Logan, 1926, p. 445). Traces of cinnabar also have been found in an area south of Shingle Springs (Logan, 1926, p. 445).

Radioactive Minerals

For the past several years there has been some prospecting for uranium in El Dorado County. A number of claims which have shown radioactivity have been located. One property has had some development work done. However, as yet no uranium ore has been mined.

Intermittent development work is being done on a prospect in the Slug Gulch area 3 miles due east of Fairplay on land owned by Clifford Smith of Somerset. This is near a hydraulic mine that was worked many years ago. Some radioactive material was found in highly weathered shear zones and in several small fissures in metamorphic and granitic rocks. Some of the soil in the area also is radioactive. It is not known if uranium is present. The area of radioactivity is several yards wide and about a quarter of a mile long. A number of shallow open cuts have been made with a bulldozer.

Claims for radioactive minerals have been located in the High Meadows area 5 miles southeast of Stateline and in an area several miles west of Silver Lake. Country rock in both these areas is granodiorite.

Silver

The recorded value of the output of silver in El Dorado County from 1880 to 1953 is $161,706. Virtually all of this was a by-product of gold and base-metal mining. Silver occurs alloyed with native gold in lode and placer deposits and with sulfide minerals in base-metal ores. At the present time, the Hazel Creek gold mine, the ore of which contain a high percentage of galena, accounts for much of the silver produced in the county.

Titanium

Small amounts of ilmenite, a titanium-bearing mineral, are found in the black sands in stream channels in El Dorado County. Ilmenite ($FeTiO_3$) is a heavy black resistant mineral that is related in origin to intrusive rocks—most commonly to basic rocks but in some cases to acidic rocks such as granite pegmatite.

Tungsten

Several thousand dollars worth of tungsten ore was mined from the Comeback Consolidated mine in 1932. However, there has been no other recorded output of tungsten ore in El Dorado County. In the past few years there has been a considerable amount of prospecting, and several prospects have been discovered. Early in 1956, the Sciaroni brothers were developing a prospect and treating small lots of ore in a pilot mill at Grizzly Flat.

Scheelite ($CaWO_4$) is found in contact metamorphic zones where calcareous rocks of the Calaveras group have been intruded by granitic rocks. The most common rock in which the scheelite occurs is tactite. Scheelite is most easily recognized by a light blue fluorescence under ultraviolet light. Sulfides, chiefly pyrite, often are associated with the scheelite.

Comeback Consolidated (Bear Creek) Mine. Location: sec. 4, T. 11 N., R. 11 E., M. D., 7 miles northeast of Placerville and 4 miles due east of Spanish Flat near the north bank of Rock Creek. Ownership: Ajax Consolidated Mines Company, c/o Bruce Thompson, First National Bank Building, Reno, Nevada.

This property has been the source of the total recorded output of tungsten ore in the county. The claims first were located in 1930 (Logan, 1938, p. 365). During 1931-32, between $3,000 and $4,000 worth of scheelite was produced from this property by sluicing (Averill, 1943, p. 318). The property was worked again in 1943 by H. G. Walker of Placerville, and some scheelite was produced by running dump material through a small jig. In the past few years the present owner has prospected the property.

The scheelite is in a calcite vein and in a fracture that intersects the calcite vein. Country rock is chlorite schist of the Calaveras group. The vein, which is 2 to 3 feet wide, strikes north-northwest and dips 50° to 60° W. Some of the scheelite occurs in masses or in coarse individual crystals. Small amounts of free gold are present. The deposit is developed by a 187-foot adit that was driven along the fracture for 100 feet and then north along the vein. Fifty feet above and below this adit an 80-foot and a 240-foot adit respectively have been driven.

Grizzly Flat Deposit. Location: sec. 32, T. 10 N., R. 13 E., and sec. 5, T. 9 N., R. 13 E., M. D., on Sturdevant Ridge 3 miles northwest of Grizzly Flat. Ownership: north portion by the Caldor Lumber Company, Diamond Springs; central portion by Horace Pierce, Sacramento, leased by E. L. Wilbur, North Sacramento; south portion, by Americo and Columbus Sciaroni, Grizzly Flat.

Scheelite was discovered in this area about 1950 (Columbus Sciaroni, personal communication, 1955). Early in 1956, the Sciaroni brothers were making a series of open cuts, and driving an adit in the south portion of the deposit. In the central portion of the deposit, E. L. Wilbur was removing the overburden by bulldozing, and samples were being taken.

Discontinuous bodies of scheelite-bearing rock occur in a northwest-trending zone of contact-metamorphic rock that is in metasediments of the Calaveras group. The zone is about 1 mile long and more than 600 feet wide. It consists of layers of garnet-epidote tactite, interbedded with quartz-mica schist, quartzite, and granite gneiss. A few small quartz veins are present. Disseminated scheelite occurs in the tactite and in the adjoining quartz-mica schist. Some scheelite was found in quartz veins near the south end of the deposit. The scheelite-bearing horizons range from 5 to 30 feet in width and are as much as several hundred feet in length. For the most part the scheelite is fine grained, although a few coarse crystals have been found. Several assays made of selected samples of scheelite-bearing tactite showed 1 to 2 percent WO_3 (Columbus Sciaroni, personal communication, 1955).

There has been no recorded production from this deposit. However, the Sciaroni brothers have been concentrating small amounts of ore in a small pilot mill and stockpiling the concentrates. The ore is crushed, fine ground, and sent through a Denver jig; the jig overflow is concentrated on a small Deister table. Concentrates are stockpiled.

Williams prospect. Location: sec. 1, T. 9 N., R. 11 E., M. D., 7 miles southeast of Placerville and 1 mile northeast of Buck's Bar. Ownership: Blanche Williams, Placerville; leased by F. T. Lee and J. Needham, Sacramento, California.

This tungsten prospect was discovered in 1954 and was in the development stage in early 1956. An open cut 20 feet long, 20 feet wide, and 15 feet high has been made on the south slope of a ridge. Several tons of ore were mined for milling tests. A number of analyses show that parts of the deposit contain from 1 to 2 percent WO_3 (F. T. Lee, personal communication, 1955).

Scheelite occurs in a northwest-trending body of garnet-epidote tactite that pinches out near the surface and widens at depth. The greatest width of exposed tactite is about 12 feet. The scheelite is in thin parallel streaks near the center of the tactite body. Quartz, calcite, and minor amounts of pyrite are present. Country rock is quartzite and mica schist of the Calaveras group. A large body of granodiorite lies to the south.

Zinc

Minor amounts of zinc are produced intermittently as a by-product of copper and gold mining. Sphalerite (ZnS), the principal zinc mineral of the Sierra Nevada, is a minor constituent of some Mother Lode and East Belt gold ores. In some portions of the foothill copper belt, however, especially to the south in Calaveras County, zinc is nearly as abundant as copper. Sphalerite is concentrated by selective flotation.

Nonmetallic Minerals

Asbestos

A total output of 142 tons of asbestos was recorded for El Dorado County during the period 1904-06. There is no record of any asbestos having been mined after 1906. Part of this asbestos was from the Contraband copper mine near Georgetown (Aubury, 1906, p. 262). Asbestos also is found at the French Hill gold mine near Spanish Dry Diggings, although there is no record of its output (Logan, 1938, p. 207).

Chrysotile and amphibole asbestos usually are found in or near massive serpentine bodies, in veins or veinlets in parallel or net-like patterns. The value of an asbestos deposit depends largely on the extent, width, and closeness of spacing of the veins. The most favorable areas for prospecting for asbestos are the serpentine belts in the western portion of the county. The largest serpentine bodies are in the Flagstaff Hill, Garden Valley, Georgetown, Latrobe, and Webber Creek areas.

Clay

Minor amounts of clay have been mined in El Dorado County. Although not shown on the geologic map, thin layers of clay are interbedded with gravels in the Tertiary river channel deposits. Channel deposits in the Georgetown, Newtown, and Placerville areas were the sources of the clay mined in the county (Aubury, 1906, p. 361).

Dimension Stone

Dimension stone was quarried in El Dorado County starting soon after the beginning of the gold rush. A number of buildings in Placerville, El Dorado, Diamond Springs and Shingle Springs, Coloma, and Lotus were constructed of rhyolite tuff, granite, and greenstone during the 1850's. Also there are many early-day stone walls and fences in the western portion of the county. Much of the stone used in these structures was quarried locally.

Rhyolite tuff of Miocene age was the most widely used stone for these early structures. Some was imported from the Volcano area of Amador County. Greenstone of the Calaveras and Amador groups, granite, and smaller amounts of serpentine, limestone, and talc schist also were used. Large quantities of boulders of andesite of Mio-Pliocene age were used to construct fences, walls, and miners' dams. Many of these were built by Chinese laborers.

In later years the demand for dimension stone for building purposes decreased. However, since 1947, rhyolite tuff has been quarried and processed in a plant in the Newtown area and marketed under the trade name of Sierra Placerite as ornamental stone for use in patios and fireplaces. Also small amounts of rough flagstone composed of rhyolite tuff have been produced intermittently in the Newtown area for the past several years.

Sierra Placerite. Location: sec. 29, T. 10 N., R. 12 E., M. D., 1 mile south of Newtown just northwest of the junction of the Newtown and

Pleasant Valley roads. Ownership: Sierra Placerite, Route 2, Box 252, Placerville, A. E. Nicolazzi, president and Warren Wilson, manager.

Ornamental building stone that is marketed under the trade name of Sierra Placerite is produced. The finished stone is used in fireplaces, patios, outdoor barbecue pits, stone facing, and garden walls. Rhyolite tuff is obtained from a nearby quarry and sawed and split into slabs and blocks. Some of it is fired in kilns. The operation was started in 1947 by R. C. Young and purchased by the present operator in 1954 (Warren Wilson, personal communication, 1955).

Extensive beds of vitric crystal rhyolite tuff of Miocene age occur in the region. In some places in this area the beds are several hundred feet thick. The tuff is fine-grained, light buff to white in color, and contains small plates of black biotite. There are numerous limonite-stained layers in the tuff, commonly in circular or concentric patterns. The quarry, which is quarter of a mile northeast of the plant, is 100 feet long, 50 feet wide, and 15 to 20 feet deep. An auger drill and power loader are used in quarrying. Large blocks of stone are trucked to the plant, while small fragments are sold directly from the quarry as rough stone. An old quarry just north of the plant was worked until 1954.

At the plant the pieces are sawed into slabs of desired thickness with circular stone saws that have carborundum or tungsten carbide teeth. The most common thicknesses are 2, 3, or $5\frac{1}{2}$ inches. The slabs are broken into desired widths, usually about 4 inches, with a hydraulic breaker or splitter. They then are trimmed to any desired length with hand axes. About two thirds of the shaped stone is marketed with natural colors. The remaining third is heated in a brick oven to drive off all contained moisture. The dried stone is then fired in an oil- and gas-fired continuous kiln. The small amount of iron present is oxidized so that the stone is colored to pleasing shades of orange, red, and pink. The lighter colors are obtained by firing at about 1600° F.; a brick-red color is obtained by firing at 2000° F. There is also a small electric kiln on the property for firing special hand-finished pieces. A ceramic glaze is put on novelty items such as desk sets and lamp bases.

Most of the finished product of the past few years has been used in fireplaces and building fronts in the numerous housing developments in the San Francisco Bay Area. The rate of production is as high as 100 tons of finished stone per month (Warren Wilson, personal communication, 1955). Seven men work at the quarry and plant.

Dolomite

A vein of dolomite occurs at the Larkin gold mine, which is 1 mile west of Diamond Springs. This vein, which is as much as 42 feet in width at the surface, is part of the Mother Lode system (Logan, 1947, p. 233). The dolomite has a rusty color because of the presence of iron oxide. Talc, quartz, and pyrite are also present.

Gems and Ornamental Minerals

About 30 small diamonds have been recovered from Tertiary channel deposits in the Smith's Flat area of El Dorado County. The largest of these weighed 1.88 carats (Murdoch and Webb, 1948, p. 130). Diamonds are believed to have originated in ultrabasic igneous rocks.

Crystallized gold in white quartz, which has been used in jewelry, has been found along the Mother Lode belt, particularly in the seam deposits between Placerville and Georgetown. It also has been found in pocket deposits in both the East and West gold belts.

Azurite and malachite occur at some of the copper mines of the foot-hill copper belt. Quartz crystals have been recovered from veins along the Mother Lode gold belt. Californite and idocrase have been found at the Cosumnes and Pioneer-Lilyama copper mines. Crystals of epidote and garnet have been found at these mines and at the Williams and Grizzly Flat tungsten prospects. Opal with various pleasing shades of brown has been found in the hydraulic mine at Slug Gulch 2 miles east of Fairplay. Occasionally fragments of opalized wood are found in the Tertiary channel deposits.

Limestone

For many years the production of limestone and lime has been a major industry in El Dorado County. At present lime and limestone are the chief mineral products of the county. The Cool-Cave Valley limestone deposit, the largest in the county, was a major source of lime-stone for a cement plant in the San Francisco area for nearly 30 years (Clark, 1954, p. 440). Four producers supply limestone for use in lime manufacture, sugar beet refining, glass manufacture, steel fluxes, roofing granules, poultry grits, mineral filler, aggregate material, and soil conditioners. Lime-manufacturing plants are active at two of the properties. Much of the finished material is marketed in the San Francisco and Sacramento-Stockton areas.

Some of the largest accessible limestone deposits of the Sierra Nevada are in the county. Many have been worked at one time or another. All of the deposits are composed of fine- to coarse-grained crystalline lime-stone that ranges from white to bluish-gray in color. On the basis of limited fossil evidence, the limestone is considered to be of Carbonifer-ous age and part of the Calaveras group. All of the deposits are lens shaped, and most have a northerly trend. They are interbedded with metasedimentary or metavolcanic rocks.

Most of the deposits utilized at present are high in calcium and low in magnesium, although one deposit is worked intermittently for mag-nesian limestone. Many are cut by dikes of fine- to medium-grained diorite, quartz diorite, and their schist equivalents. Also, some of the limestone deposits contain veins of chert. Jointing is prominent in much of the limestone.

The limestone deposits in the county are mined by open-pit and underground methods. In underground mining, shrinkage stopes are employed. At one time glory holes were employed by one producer. At various times one operator produces an entire year's supply with one blast. One producer delivers stone from the quarry to the lime plant via a 3-mile aerial tramway. Finished material from the plants is shipped by both truck and railroad.

Cool-Cave Valley (Coswell-Cave Valley) Deposit. Location: secs. 6, 7, and 18, T. 12 N., R. 9 E., M. D., 4 miles east of Auburn on the south side of the Middle Fork of the American River. Ownership: South ⅔, Henry Cowell Lime and Cement Company, 2 Market Street, San Fran-cisco; north ⅓, Ideal Cement Company, 310 Sansome Street, San Francisco.

FIGURE 23. Limestone quarry of California Rock and Gravel Company.
Camera facing north.

The largest limestone deposit in the county, the Cool-Cave Valley deposit has been a major source of limestone for the cement, lime, and beet-sugar industries for many years. It consists of two north-trending lenses. During the 1880's and 1890's, limestone was quarried from the south lens and burned in stone lime kilns (Clark, 1954, p. 441). Ruins of these old kilns are near State Highway 49. From 1910 to 1940, the Pacific Portland Cement Company operated an extensive quarry at the north end of the north lens by the Middle Fork of the American River. This deep quarry was known as Mountain Quarries. Limestone was sent through a crushing and sizing plant and shipped over a company-owned railroad to Auburn and then to the Pacific Portland Cement plant in Solano County or to beet sugar refineries. In 1942, the quarry was abandoned and the railroad dismantled.

In 1946, the south portion of the north lens, south of Mountain Quarries, was leased by the California Rock and Gravel Company. This concern produces limestone for use in beet-sugar refining and lime manufacturing. A detailed study recently was made of this deposit, the results of which were published in the California Journal of Mines and Geology, vol. 50, July-October 1954, pp. 439-465.

The deposit consists of high-calcium bluish-gray crystalline limestone. The north lens, which is cut by the Middle Fork of the American River near the north end, is about 5500 feet long and has an average width of 400 feet. It was worked to a depth of nearly 800 feet in Mountain Quarries. The south lens, which is crossed by State Highway 49, is 2,000 feet long and nearly 600 feet wide in the middle. A number of samples taken from both lenses averaged more than 97 percent $CaCO_3$ and less than 1 percent $MgCO_3$ (Clark, 1954, p. 454). Country rock is green schist and massive greenstone. Both lenses are cut by dikes of medium-grained diorite and quartz diorite. The limestone is well jointed.

At the California Rock and Gravel Company quarry in the south-central portion of the north lens, limestone is mined by several methods. In 1946 and 1947, it was quarried by benching with 30-foot wagon drill holes. From 1948 to 1952, it was mined with "coyote" holes where adits were driven perpendicular to the quarry face and branches driven perpendicular to the adits. These were loaded with dynamite and an entire year's supply of limestone was dislodged in one blast (Clark, 1954, p. 458). At the present time large benches and 100-foot vertical wagon drill holes are used. About six months' supply is dislodged in one blast. Large fragments remaining from the blast are broken with a drop ball. In the fall of 1955, a new quarry north of the plant was being developed.

Broken stone from the quarry is trucked to a nearby crushing and screening plant. Coarse and medium sizes are trucked to Auburn and shipped by rail to beet-sugar refineries in the Sacramento and San Joaquin Valleys. The finer sizes are trucked to the Diamond Springs Lime Company for lime manufacturing. Until 1954, this concern also supplied limestone to the Vertin Lime Company plant at Rattlesnake Bridge. Undersize is stockpiled and sold as road metal. Twenty-seven men work at the quarry and plant.

Another limestone lens is located about one mile to the west on the Middle Fork of the American River. It also is composed of bluish-gray high-calcium limestone (Clark, 1954, p. 460). The lens, which strikes northwest, is 450 feet long and 50 feet wide. It was worked many years ago, and the stone was burned in a stone lime kiln at the north end.

Diamond Springs Lime Corporation. Location: the plant is in sec. 24, T. 10 N., R. 10 E., M. D., at Diamond Springs; the quarry is in sec. 28, T. 10 N., R. 11 E., M. D., 3 miles to the east. Ownership: Diamond Springs Lime Corporation, Diamond Springs, H. C. Green, manager.

This company operates a lime plant that was erected in 1927 (Logan, 1938, p. 274). Various grades of lime and quicklime are produced for use in steel mills, the building industry, and agriculture. Aggregate material, road metal, and asphalt mix are by-products.

Most of the limestone used in the plant is imported from El Dorado Limestone Company at Shingle Springs and the California Rock and Gravel Company at Cool. Limestone from these two concerns is shipped in by truck or rail and dumped into an underground bunker. From here, it is delivered by clamshell bucket to a storage shed, then screened. As the size required for the kilns averages $\frac{3}{4}$ of an inch to $1\frac{1}{2}$ inches, oversize stone is sent by belt conveyor to a jaw crusher. Undersize is stockpiled and sold as aggregate or is sent to a hot plant on the property for use in the manufacture of asphalt mix.

From the storage shed, the limestone is belt-conveyed to bins which supply the kilns. From the bins, the limestone goes by belt feeders to two oil-fired 8- by 125-foot inclined rotary kilns. As the limestone from the two suppliers have different physical characteristics, only one type is burned at a time (H. C. Green, personal communication, 1955). A precipitator at the head of the conveyor belt collects much of the dust.

Lime from the kilns goes through two 4- by 48-foot coolers under the kilns and then by pan conveyor and bucket elevator to two bins. From these bins it is sent by a screw conveyor and more bucket ele-

FIGURE 24.   Diamond Springs Lime company plant.

vators either to a hydrator or to a rolling mill for fine grinding. Some of the lime is sent to a recently installed unit consisting of a slaker, drier, and tube mill for the production of "Marvel," a trade name for a high-calcium lime of minus 100-mesh size. In the summer, when the company-owned quarry is in operation, 7 percent magnesian lime is produced.

The various finished products are sent to storage bins. Material from the storage bins is bagged or bulk-loaded into railroad cars and trucks for shipment. About half of the finished product is sold to steel plants in the San Francisco Bay area as fluxing material (H. C. Green, personal communication, 1955). Another major consumer is the building industry, in which lime is used for manufacturing plaster, stucco, and mortar. Smaller amounts are used in agriculture and the chemical industry.

The limestone deposit, 3 miles east of the plant, is mined during the summer. A limestone lens, which is about 2,500 feet long and as much as 500 feet wide, occurs in metasedimentary rocks of the Calaveras group. It strikes north and has a vertical dip. Jointing is common. The limestone is fine-grained and ranges from white to bluish-gray in color.

The stone is quarried in 30-foot benches with wagon drills and jackhammers. The broken stone is loaded by power shovel into dump trucks and delivered to a primary jaw crusher. From the jaw crusher, it goes by conveyor belt to storage bins that serve the aerial tramway. This tramway, 3 miles long, uses 149 buckets of 800 pounds capacity each and can deliver as much as 30 tons an hour to the plant.

El Dorado Limestone Company. Location: sec. 15 and 22, T. 9 N., R. 9 E., M. D., 3 miles southwest of Shingle Springs. Ownership: El Dorado Limestone Company, Shingle Springs, California, J. H. Bell, president; C. R. Nichols, general manager; and F. G. DeBerry, mine superintendent.

The El Dorado Limestone Company produces high-calcium limestone for various uses. Finished material is used in the manufacture of lime, by steel mills, glass manufacturers, in beet-sugar refining, and in construction materials.

The present concern, which was formed in 1931, is a successor to El Dorado Lime and Minerals Company, which began mining the deposit in 1918 (Logan, 1947, p. 230). Prior to that date, limestone was

FIGURE 25.   Limestone quarry of the Diamond Springs Lime company.
Camera facing west.

quarried just north of the mine. This was burned in nearby stone lime kilns and the lime used for building purposes. Ruins of these old lime kilns may still be seen.

This deposit consists of two parallel northwest-trending lenses of crystalline limestone which are part of a belt of metasedimentary rocks of the Calaveras group. Dip is vertical or nearly so. The two lenses are from a few to more than 50 feet apart. Country rock is slate and schist. The east limestone lens has an average width of 60 feet; the west lens has an average width of 40 feet. Several dikes of fine- to medium-grained granodiorite cut both limestone lenses. Because of the thick soil overburden, outcrop, of the deposit are poor. It is shown on the U. S. Geological Survey Placerville folio as being extremely limited in size (Lindgren and Turner, 1894).

The limestone has an even, medium- to coarse-grained texture. It is mostly grayish-white in color, although some gray limestone is present. Jointing is prominent, the principal joint planes dipping both north and south. All of the limestone is high in calcium and low in silica and magnesium. Sixty-seven carloads shipped to one user averaged 97.65 percent $CaCO_3$ and 0.24 percent SiO_2 (Logan, 1947, p. 231).

The main working entry is a 1,000-foot three-compartment vertical shaft sunk near the east wall of the east lens. The deepest workings are on the 800-foot level. Crosscuts extend from the shaft stations to the west lens. No timbering is required. Good ventilation is facilitated by a raise connecting the 150-foot level with the surface south of the shaft,

FIGURE 26.   El Dorado Limestone Company—headframe with hoist house to right, crushing and sizing plant to left. Camera facing east.

a raise from this level to the old quarry north of the shaft, and numerous raises that connect the lower levels.

For many years shrinkage stopes have been employed in the mine. The main haulageways are drifts about 20 feet wide and 8 feet high that have been driven north and south from the shaft stations in both the east and west lenses. In developing a stope, short crosscuts are driven perpendicularly from the main haulageways; the ends of these are used as drawpoints and are connected by ore chutes to the bottoms of the stopes at the sides. The bottoms of the stopes have an inverted V-shape so the loose rock flows down ore chutes to the drawpoints. The lowest parts of the stopes are about 12 feet above the level of the roofs of the drifts.

Gardner Denver drifters and liners and 30 percent ammonia dynamite are used for drilling and blasting. The rock is drilled and blasted in 5-foot slices in the stopes. Broken rock is drawn from the ore chutes and is allowed to pile up at the end of the crosscuts. Track-mounted Eimco mucking machines in the short crosscuts load the broken rock into a 3½-ton ore car in the haulageway. Occasionally large chunks are blasted in the haulageways. The rock is delivered to the loading pockets at the shaft station where the larger pieces are broken on the grizzlies with sledge hammers.

Limestone is mined in stopes in both lenses on the 650-foot level. North and south drifts are being driven on the 800-foot level, preparatory to the development of new stopes between that level and the 650-foot level. When the stopes have been mined up to the level above,

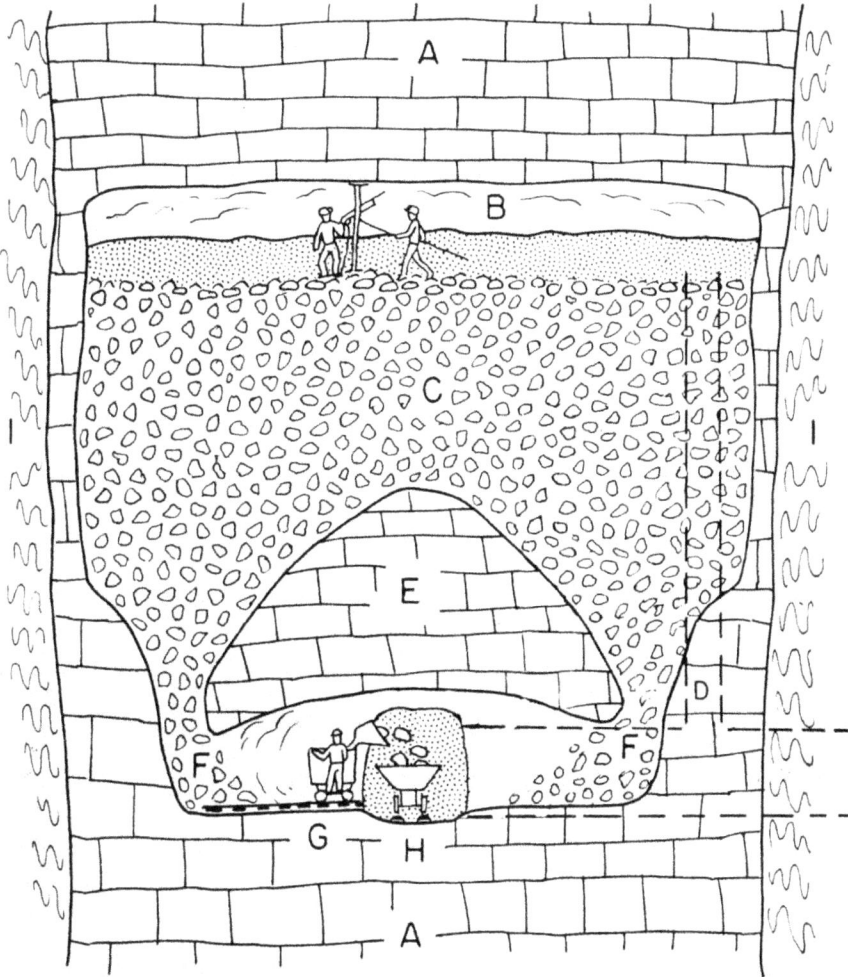

A	-	Limestone lens
B	-	Five foot slice
C	-	Broken limestone
D	-	Raise
E	-	V-shape base of stope
F	-	Drawpoint
G	-	Automatic loader
H	-	Haulageway
I	-	Slate and schist wall rock

FIGURE 27.   Cross-section of shrinkage stope at El Dorado
Limestone Company limestone mine.

the remaining pillars on that level also are mined out. Completed stopes reach maximum dimensions of 300 feet in height, 600 feet in length, and 70 feet in width. The limestone is mined to within a few feet of the wall rock.

Limestone is hoisted in 2-ton skips and deposited in a 200-ton silo-type bin mounted on the 100-foot steel headframe. From the bin, the rock passes by gravity through a Kue-Ken crusher, vibrating screens, and trommels. During this process, the material is hand picked on conveyor belts to remove fragments containing schist or dike rock. Sized material is delivered to separate bins from which it can be loaded into railroad cars or trucks. Some of the coarser material is stacked in yard stockpiles. The company has 1.9 miles of spur track which connects with the Southern Pacific Railroad.

The most important of the finished products are stone used in the manufacture of lime, roofing gravel, material used in glass-making plants, and sugar rock. The stone used in lime manufacturing is shipped to the Diamond Springs Lime Company, Diamond Springs. Sugar rock is sent to beet-sugar refineries in the Sacramento-Stockton area. Other products are fluxing material for open-hearth steel furnaces and filler material used in the manufacture of paints and linoleum. Agricultural limestone also is produced.

Marble Valley (Schwalin) Deposit. Location: secs. 8 and 17, T. 9 N., R. 9 E., M. D., 2 miles southeast of Clarksville in Marble Valley. Own-

FIGURE 28.   Semon Lime Company plant near Rattlesnake Bridge.
Camera facing south.

ership: Henry Cowell Lime and Cement Company, 2 Market Street, San Francisco, California.

Years ago, limestone was quarried at this deposit and burned in a vertical kiln a quarter of a mile to the west. The last period of operation was about 1918 (Henry Cowell Lime and Cement Company, personal communication, 1955).

Limestone crops out at intervals for a distance of nearly 4,000 feet, but the total extent of the deposit is unknown. The deposit ranges from less than 100 to 200 feet in width. It apparently is lensoid and has a north-northwest strike. The limestone is medium grained and ranges from white to light bluish-gray in color. An analysis of a composite sample taken by C. A. Logan is as follows (Logan, 1947, p. 232):

	Percent
Insoluble	0.43
Ferric and aluminum oxide	0.42
Calcium carbonate	98.80
Magnesium carbonate	0.30
	99.95

The deposit is developed by two open quarries. The north quarry is about 150 feet wide and several hundred feet long with quarry faces up to 40 feet high. The south quarry is 180 feet wide, 300 feet long, and 20 to 25 feet deep.

Rattlesnake Bridge (Alabaster Cave, Rattlesnake Bar) Deposit. Location: sec. 15, T. 11 N., R. 8 E., M. D., 1 mile east of Rattlesnake Bridge and 5 miles south of Auburn. Ownership: Semon Lime Company, Auburn, H. S. Dahlman, manager.

This limestone deposit, which also has been known as the Alabaster Cave deposit, has been worked intermittently since the 1860's by a number of concerns. There is an old stone lime kiln near the Rattlesnake Bridge road, which crosses the northern portion of the deposit. From 1930 to 1942, the Auburn Chemical Lime Company worked the deposit. Limestone from the south end of the deposit was treated at a crushing and sizing plant and sent through a lime plant. Much of the lime manufactured during the 1930's was sold to gold mines for use in cyanidation (Logan, 1947, p. 223).

From 1942 to 1954 there was little output from the quarries, as most of the limestone used at the plant was purchased from the California Rock and Gravel Company at Cool. From 1946 to 1948, the plant was operated by the Auburn Lime Products Company; from 1949 to 1954 by the Hughes-Vertin Lime Company; and from 1954 to 1955 by the Vertin Lime Company. In July 1955, the Semon Lime Company purchased the holdings and has operated the property since (H. S. Dahlman, personal communication, 1955). A new 7- x 110-foot rotary kiln and increased storage facilities were installed in 1953. In 1954, a new quarry was opened in the north end of the deposit. At present several grades of lime, quicklime, and hydrated lime are manufactured and marketed under the trade name "Versatile." Sized limestone also is sold.

The deposit is a lens of coarse-grained crystalline limestone ranging from white to bluish-gray in color. It strikes north and dips steeply east. It is about 4,000 feet long and 80 to 100 feet wide. Parallel joints are prominent; the principal joint planes strike west and dip steepl south. Most of the limestone is high in calcium carbonate (97 to 9 per cent) (H. S. Dahlman, personal communication, 1955). At one tim there were a number of natural caves, which since have been quarrie out. Country rock is amphibolite with small amounts of serpentin There are several bands of amphibolite and chlorite schist up to severa feet thick in the limestone, and these are troublesome in quarryin operations.

The deposit is developed by four quarries. Two are located sout of the plant: one, 300 feet long, extends north from the south end c the deposit and the other 475 feet long extends south from the plan They are up to 75 feet deep and 25 to 50 feet wide. These were the mai source of limestone prior to World War II. An old quarry immediatel north of the road was the source of rock from the old stone lime kil

Limestone is mined from 20-foot benches in a quarry about 300 fe long and 25 to 50 feet wide in the north end of the deposit. Wage drills with 40 percent gelatin dynamite are used; large boulders a broken with jackhammers. Primary crushing is done in the quarr and secondary crushing and screening at the plant. The plus $\frac{1}{4}$-inc stone is sent either to the new 7- by 110-foot oil-fired inclined rotar kiln or two older 4- by 60-foot kilns. Lime passes through a rotar cooler and into a storage bin for sacking, bulk loading, or hydratio Most of the lime produced is shipped to steel plants in the San Fra cisco Bay area or to the building-trades industry in the Stockton-Sa ramento area. Minus $\frac{1}{4}$-inch rock is screened into several sizes and so as roofing granules, chicken grits, limestone flour, and road metal. Fi teen men work at the plant and quarry.

Slug Gulch (Cosumnes) Deposit. Location: secs. 24, 25, and 26, 9 N., R. 12 E., M. D., by the Middle Fork of the Cosumnes River, miles northeast of Fairplay. Ownership: Clifford Smith, Somerset.

This limestone deposit crops out for a distance of $1\frac{1}{2}$ miles in a nort east direction from Slug Gulch to north of Rocky Bar on the Cosumn River. It is lensoid and has an average width of several hundred fee The limestone is medium- to coarse-grained in texture and ranges fro white to bluish-gray in color. Other than the quarrying of a sma amount of limestone at the north end for use as road metal, there ha been no work on the deposit (Lee Miller, personal communicatio 1955). The deposit is accessible by several dirt roads, one of which e tends east from Fairplay to the Cosumnes copper mine, which is ju west of the deposit.

Mineral Paint

Although the production of mineral paint has never been recorde from El Dorado County, samples of both red and yellow ochre we submitted to the Division of Mines Laboratory many years ago (Loga 1927, p. 444).

Silica

Large quantities of quartz occur in El Dorado County. There are n only numerous massive quartz veins in the county, but quartz is major constituent of many of the other rocks, particularly quartzite ar granitic rocks. Also, quartz pebbles and cobbles are common in Te iary and Recent stream gravels. Quartz has been mined from three properties in the county (see tabulated list under Silica).

Pure silica is used in glass-making, in glazes for porcelain and pottery, as fluxing material in foundries and smelting, in the manufacture of ferro-silicon for the steel industry, and in abrasives. Much of the raw material for these industries must be beneficiated. Although the majority of silica for these uses is obtained from other sources, the massive quartz veins, quartzite bodies, and quartzitic gravels of the Sierra Nevada exist as potential sources of silica.

Slate

Slate has been mined in El Dorado County since about 1887. The Placerville area has been the largest and most consistent source of slate in California. At present most of the slate mined in California is from El Dorado County.

Nearly all of the slate mines in the county are between Placerville and Kelsey, where the Mariposa slate makes up a northwest-trending belt that averages 2 miles in width. Dip is vertical or steeply to the northeast. The slate is bluish-black in color when fresh and weathers to a brown or greenish-gray. Much of it is in uniform beds, with only a few small quartz veinlets or minor amounts of pyrite as impurities.

For many years, most of the slate mined in the county was dimension slate or squares, a unit equivalent of 100 square feet. The peak in slate production was reached in 1906 when 10,000 squares valued at $100,000 were produced. Dimension slate was used principally in shingles. Smaller amounts were used for blackboards, school slates, sinks, and as finished stone in homes. Major sources of dimension slate were the El Dorado Slate Products Company, which operated a quarry on the south side of Big Canyon 1½ miles north of Placerville, and the Eureka Slate Company, which owned a quarry 1 mile south of Kelsey. Slate from the Eureka quarry was delivered to Placerville via a 13,000-foot aerial cableway, which crossed the American River at an elevation of 600 feet above the bed of the river.

Because of the great amount of hand labor required and the resulting high costs, the production of dimension slate in El Dorado County gradually declined during the 1920's and finally ceased altogether. However, about this time the demand for crushed and ground slate for use as roofing granules and as filler material increased. In 1928, the old Chili Bar quarry was reopened by the Pacific Minerals Company. At present, this is the only active slate property in the county.

Chili Bar Mine. Location: sec. 36, T. 11 N., R. 10 E., M. D., on the south side of the South Fork of the American River just east of the Chili Bar bridge, 3½ miles north of Placerville. Ownership: Pacific Minerals Company, Ltd., 339 10th Street, Richmond, California; G. H. Bishop, manager; Edward Bishop, foreman.

Originally an open-pit operation, the Chili Bar slate mine first was worked from 1887 to 1897 (Logan, 1926, p. 448). In these early operations, roofing shingles and other forms of dimension slate were produced. The property was idle until 1928, when it was reopened by the present operator. An underground mine was developed and a

FIGURE 29. Pacific Minerals Company slate mine and crushing plant at Chili Bar, by South Fork American River. Camera facing south.

grinding and sizing plant installed (G. H. Bishop, personal communication, 1954). Since that date, roofing granules and slate-dust filler have been the principal products with the exception of minor amounts of dimension slate.

Dark fissile slate of the Mariposa formation occurs in a northwest-trending belt about 2 miles wide. The slaty cleavage strikes N. 10° W. and dips 85° NE. Metasedimentary rocks of the Calaveras group lie east of this belt and fine-grained metadiabase is exposed to the west. The mine is about a quarter of a mile east of the west boundary of the slate. Small quartz stringers mostly less than an inch wide are present here and there in the slate. Several hundred yards east of the mine are several quartz stringers that contain pyrite and some gold.

The mine is worked through a southeast-trending drift adit which is the main haulageway. Until 1953, the slate was mined in shrinkage stopes. These were abandoned, and at present the slate is mined by overhand stoping in pillar and chamber workings. A stope is developed by driving a drift for about 150 feet; this is then widened to 40 to 50 feet and heightened to about 18 feet. Broken slate is removed for convenient access, and full-width overhand slices, each about 18 feet high, are removed from the back of the long stope, the miners always standing on the broken slate. The stopes are finally completed leaving large open chambers with pillars between. The slate has been mined for distances of 500 feet east and west of the main haulageway, about 800 feet southeast of the portal. Little or no timber is used in the underground workings. Ingersoll-Rand and Chicago pneumatic long-shell drifters with 4-foot changes in steel and ''carset'' bits are used.

A round usually consists of about 200 60-foot holes in an 18- by 40-foot face. "Geladine" ready-split stick powder with a millisecond delay firing system is used in blasting.

Broken slate is loaded into ore cars by pneumatic mucking machines and hand-trammed to the raw-feed hopper 100 feet in from the adit portal. The slate is sent through a 2-inch grizzly. At the plant, which is just west of the adit portal, the minus 2-inch material falls into a rotary drier, while plus 2-inch slate goes to two Williams Jumbo Junior hammer mills. The crushed slate is sent through four Tyler Hum-mer double-deck vibrating screens. Plus 10-mesh material is returned to the circuit for re-crushing. Minus 10- plus 35-mesh material is treated with a special saturating oil and bagged for shipment as roofing granules. Minus 35-mesh material is further ground in a Joshua Hendy ball mill, elevated, and air classified in a Sturdevant air classifier. Two nominal sizes (200-mesh and 325-mesh) are prepared, bagged in paper sacks, and shipped to the market as slate-dust filler. Plant output is about 40 percent granules and 60 percent filler dust (Edward Bishop, personal communication, 1955).

Ten are employed at the mine and mill. The mine operates on one shift and the mill on two shifts per day. Bagged material is trucked to Placerville for rail shipment. Much of the finished product is shipped to the San Francisco and Los Angeles areas, and some is shipped to Oregon.

Soapstone

For many years varying amounts of soapstone have been mined in El Dorado County. Soapstone is mined at the Swift property by the Pacific Minerals Company and intermittently at the Bernett and Hayden properties. It is marketed in the San Francisco Bay area for use as a carrier in insecticide dust.

Sawed slabs and blocks of soapstone were produced as early as the 1880's in El Dorado County (Logan, 1926, p. 450). These were used as building materials and for tubs, sinks, and stoves. Later substantial quantities of soapstone were mined for use as filler material for rubber, cloth, paint, paper, roofing and as a lubricant and in electrical insulation. Sources of soapstone during these times were the Swift mine near Latrobe, which operated from 1916-24, and the Rossi or Shingle Springs mine south of Shingle Springs, which was active from 1919-28. The Swift mine was reopened in 1928 by the Pacific Minerals Company and has been continuously active since.

Soapstone deposits in the Sierra Nevada are in or near talc-rich amphibolite schist or serpentine bodies in the western foothills. Most of the soapstone that has been mined in El Dorado County has been from the Latrobe-Shingle Springs area where there are extensive bodies of serpentine. The deposits range from small lenses a few inches thick to those hundreds of feet in extent. The soapstone is most commonly greenish-gray in color. Because of the iron oxide present in most Sierran soapstone, it has not been used in ceramics.

Bernett Property. Location: sec. 23, T. 9 N., R. 9 E., M. D., 4 miles southwest of Shingle Springs adjacent to the Southern Pacific Railroad. Ownership: D. P. Bernett, Shingle Springs, California; leased by Frank Harris, 580 North L Street, Livermore, California.

Soapstone has been mined intermittently at this property by Mr. Harris since 1953. It is mined by stripping with a bulldozer. The soapstone is trucked in crude form to the Industrial Minerals and Chemical Company in Berkeley, where it is ground for use as an insecticide carrier.

The deposit is a north-trending body of talc schist at least 500 feet in length. Width of the body ranges from 40 to 60 feet, and it has been bulldozed to a depth of 10 feet. The schist is fine grained, and its color ranges from light gray to reddish brown, the latter caused by iron oxide staining. The presence of iron oxide does not lower its value as an insecticide carrier (Frank Harris, personal communication, 1954). The pit is approximately 100 feet long and ranges from 40 to 60 feet in width.

Hayden Property. Location: sec. 7, T. 9 N., R. 10 E., M. D., 1¾ miles southwest of Shingle Springs. Ownership: Mark Hayden, Shingle Springs; leased to Industrial Minerals and Chemicals Company of Berkeley.

Since early 1955 the Industrial Minerals and Chemical Company has mined soapstone intermittently from this property. It is trucked to the company's grinding plant at Florin, Sacramento County, and prepared for use in insecticides.

The deposit is a lensoid body of iron oxide-stained greenish soapstone which is on the west side of a small body of metasedimentary rock enclosed in greenstone. The soapstone is quarried from an open cut which is about 100 feet long, 40 feet wide, with a 20-foot face at the north end. A jackhammer is used in drilling. The excavated material is loaded into trucks with a Caterpillar loader. Frank Harris of Livermore has the contract to mine and haul the ore to plant.

Pacific Minerals (Swift) Mine. Location: sec. 35, T. 9 N., R. 9 E., M. D., 3 miles northeast of Latrobe. Ownership: Rufus Swift, Shingle Springs; leased by Pacific Minerals Company, Ltd., 337 10th Street, Richmond, California; George McKenzie, mine foreman.

This property was worked originally prior to 1920 (Logan, 1920, p. 433). It was active again during the early 1920's. In these early operations, soapstone was mined in underground workings by hand labor, using augers and drills. The soapstone was shipped to San Francisco where it was used for coating in prepared roofing. The workings caved around 1924, and the mine was shut down (Logan, 1926, p. 451). In 1928, the property was reopened by the present operator and has been worked continuously since that date (George McKenzie, personal communication, 1955).

Until recently, the operation was seasonal, generally running from March through September, but now it is a year-around operation. The crude soapstone is shipped by rail to grinding mills in the San Francisco Bay area where it is prepared for use in insecticides.

The soapstone forms in a lensoid body that is as much as 100 feet wide. It strikes north-northwest and dips moderately to steeply northeast. The soapstone is greenish-blue in color, although much is stained with iron oxide. Country rock is slaty greenstone and metasedimentary rocks.

From 1928 to the late 1930's the mine was worked by underground methods. During that time the main entry was a 220-foot west-trending crosscut adit. About 60 feet in from the portal the main drifts were driven 120 feet north and 200 feet south. Two other drifts, one east and the other west of the main drifts and each 200 feet long, were driven from the main drifts.

In the latter 1930's, the central portion of the mine caved (George McKenzie, personal communication, 1955). Since then soapstone has been mined from an open pit. The main pit is 175 feet long, 35 to 70 feet wide, and has a 75-foot face at the north end. There are two smaller pits at the north end of the deposit.

A jackhammer powered by a 210-cubic-foot compressor is used for drilling. The ore is blasted and loaded into a 5-ton dump truck by a D-4 Caterpillar tractor-loader. The soapstone is trucked about 800 feet and deposited into loading bins by the tracks of the Southern Pacific Railroad. A concrete stockpiling ramp, which can accommodate 2000 tons of ore, was completed in 1955. Three men work at the mine.

Stone

Besides dimension stone and slate, which have been described in separate articles, El Dorado County is a source of sand and gravel, crushed rock, riprap, and road metal. Most of the output is used locally.

The Harms Brothers Construction Company of Sacramento produces crushed rock at a plant just south of Placerville which is used as asphalt mix. Sand, gravel, and crushed rock are produced by the Cold Springs Sand and Gravel Company from Webber Creek 4 miles west of Placerville. H. C. Anderson produces sand derived from decomposed granite and gravel from the Upper Truckee River; these are used in a batching plant at Meyers for ready-mix concrete. Sand and gravel also are produced by Archibald Butler of Meyers from the Upper Truckee River.

Sand and gravel deposits largely consist of accumulations of sand, pebbles, cobbles, and boulders that have been deposited by stream action. They are poorly sorted and are elongated in the direction of stream flow. These deposits are derived from the more weather-resistant rocks and are composed of chert, metachert, quartz, quartzite, greenstone, and granitic rocks. Sand and gravel deposits are found in stream channels of both Recent and Tertiary age. Most are deficient in the finer sizes. The most extensive deposits in the county are near Placerville, where large quantities of gravel have accumulated from hydraulic mining, along the South Fork of the American River in the Coloma-Lotus area, and along the Middle Fork of the American River east of Auburn.

Coarse quartzitic sand derived from decomposed granite and granodiorite is recovered by the California Division of Highways from an extensive pit 1 mile northwest of Coloma by State Highway 49. The sand is used both in highway repair work and in asphalt mix. Several other smaller sand pits intermittently operated by the state are on U. S. Highway 50 between Placerville and Echo Summit and just south of Tahoe Valley. The El Dorado County road department mines coarse sand from a number of small open pits in the western portion of the county, particularly in the region northwest of Shingle Springs.

FIGURE 30. State Division of Highways sand pit, in deeply weathered granite in Coloma area. Camera facing east.

Waste limestone from the plants of the California Rock and Gravel Company at Cool and the Diamond Springs Lime Company at Diamond Springs is sold as road metal and fill. Serpentine, which is used by the county road department as road metal, is quarried at the Hummingbird Ranch 1 mile southwest of Garden Valley. Another smaller open pit in serpentine has been worked just north of Four Corners.

Large quantities of broken granodiorite have been used as riprap in the rebuilding and repair of U. S. Highway 50 over Echo Summit, as well as in repair work on State Highways 49, 88, and 89. Considerable quantities of broken metachert and quartzite of the Calaveras group were quarried recently at Hazel Valley for use in the construction of the Sly Park Dam, a combination earth and rock-fill dam.

Cold Springs Sand and Gravel Company. Location: sec. 9, T. 10 N., R. 10 E., M. D., on Webber Creek, 4 miles west of Placerville. Ownership: Jerry Brown, Placerville; leased to L. D. Forni of Placerville.

From 1950 to April 1953, sand, gravel, and crushed rock were produced from this deposit by El Dorado Rock and Sand Company. Since April 1953, the property has been leased by L. D. Forni who now operates it under the name of Cold Springs Sand and Gravel Company (L. D. Forni, personal communication, 1955). A wide variety of aggregate that is marketed in the Placerville area is produced. The output is as much as 300 tons of finished material per day. Since August 1955, waste rock from the Sierra Placerite operation (see Dimension Stone herein) in the Newton area has been prepared for roofing rock.

Sand and gravel are excavated from the banks of Webber Creek by a dragline mounted on a truck. An end-dump truck hauls the excavated material a quarter of a mile to the aggregate plant. At the plant, the material falls through a 9-inch grizzly and belt conveyed to the screening tower. Oversize is sent through a primary jaw crusher and a secondary gyratory crusher and returned to the circuit. A sand drag is used for sand production. Products from the plant fall into two elevated bins from which trucks are loaded. Considerable amounts of aggregate are stockpiled in the yard.

Waste rock from the Sierra Placerite operation consists of buff to red-colored rhyolite tuff. This is crushed to a ¾-inch size in the plant and loaded into 80-pound sacks. This material is marketed as roofing rock.

FIGURE 31.   Excavating gravel at Harms pit in Coon Hollow.
Camera facing south.

Harms Pit. Location: sec. 18, T. 10 N., R. 11 E., M. D., in Coon Hollow, 1 mile south of Placerville. Ownership: Harms Brothers Construction Company, Sacramento, Wally Basanko, foreman.

Since early 1955 the Harms Brothers Construction Company of Sacramento have operated a portable hot-mix plant. The output is used for surfacing the new U. S. Highway 50 bypass in Placerville. Plant capacity is about 100 tons per hour (Wally Basanko, personal communication, 1955).

Andesitic detritus and some hydraulic mine tailings containing quartz, granite, and chert pebbles and cobbles are excavated by a bulldozer. Excavated material is belt-conveyed to an Austin Western portable crushing and sizing plant. Two products are stockpiled by belt conveyors adjacent to the plant, minus $\frac{7}{16}$-inch and plus $\frac{7}{16}$- minus ¾-inch material. A Northwest power shovel with a clamshell bucket deposits the graded material in kiln-receiving bins. From here the material is draw-blended on a belt conveyor and dumped into an oil-fired kiln. In the kiln the aggregate is heated to 300° F. A bucket elevator deposits the kiln product into an elevated surge bin. From the surge bin the aggregate is dropped into an elevated pug mill along with pre-heated oil. From the pug mill, the hot mix is loaded directly into end-dump trucks and hauled to the job site.

BIBLIOGRAPHY

Allen, J. E., 1941, Geological investigation of the chromite deposits of California: California Div. Mines Rept. 37, pp. 135-139.

Armstrong, J. F., 1902, Register of mines and minerals, El Dorado County, California Min. Bur., 17 pp., 2 maps.

Aubury, L. E., 1902, The copper resources of California: California Min. Bur. Bull. 23, pp. 174-182.

Aubury, L. E., 1903, Quicksilver resources of California: California Min. Bur. Bull. 27, p. 190.

Aubury, L. E., 1906, The structural and industrial materials of California: California Min. Bur. Bull. 38, pp. 67-68, 150-152, 262, 297, 360-378.

Aubury, L. E., 1908, The copper resources of California: California Min. Bur. Bull. 50, pp. 210-220.

Averill, Charles V., 1943, Current notes on activity in the strategic minerals, Sacramento field district: California Div. Mines Rept. 39, pp. 71-76 [El Dorado County, p. 72] . . . pp. 139-141 [El Dorado County, p. 140] . . . pp. 311-322 [El Dorado County, pp. 317-318] . . . pp. 551-559 [El Dorado County, pp. 556-557].

Averill, Charles V., 1946, Placer mining for gold in California: California Div. Mines Bull. 135, pp. 255-257.

Bedford, R. H., 1946, Exploration of the El Dorado County copper mine, El Dorado County, California: U. S. Bur. Mines Rept. Inv. 3896.

Boalich, E. S., 1922, Notes on iron ore occurrences in California: California Min. Bur. Rept. 18, p. 110.

Bowen, O. E. Jr., and Crippen, R. A. Jr., 1948, Geologic maps and notes along Highway 49: California Div. Mines Bull. 141, pp. 35-86.

Bradley, W. W., 1918, Quicksilver resources of California: California Min. Bur. Bull. 78, p. 42.

Bradley, W. W., Huguenin, Emile, Logan, C. A., Tucker, W. B., and Waring, C. A., 1918, Manganese and chromium in California: California Min. Bur. Bull. 76, pp. 131-144.

Cater, F. W. Jr., Rynerson, G. A., and Dow, D. H., 1951, Chromite deposits of El Dorado County, California: California Div. Mines. Bull. 134, pt. 3, chap. 4, 167 pp.

Clark, W. B., 1954, The Cool-Cave Valley limestone deposits, El Dorado and Placer Counties, California: California Div. Mines Rept. 50, pp. 438-465.

Cox, M. W., Wyant, W. C., and Heyl, G. R., 1948, Geology of the Lilyama and Pioneer mines, El Dorado County: California Div. Mines Bull. 144, pp. 43-47.

Crawford, J. J., 1894, Mines and mining products of California: California Min. Bur. Rept. 12, pp. 21-411 [El Dorado County, pp. 101-127, 359, 391, 400-402].

Crawford, J. J., 1896, Thirteenth report (third biennial) of the State Mineralogist for the two years ending September 15, 1896: California Min. Bur. Rept. 13, 726 pp., illus. [El Dorado County, pp. 58, 131-165, 528-531, 628, 642].

Cutter, D. C., 1948, The discovery of gold in California: California Div. Mines Bull. 141, pp. 13-17.

Davis, W. M., 1948, The lakes of California: California Div. Mines Rept. 44, pp. 201-242.

DeGroot, Henry, 1882, Hydraulic and drift mining: California Min. Bur. Rept. 2, app., pp. 131-190 [El Dorado County, p. 188].

DeGroot, Henry, 1890, El Dorado County: California Min. Bur. Rept. 10, pp. 169-182.

Eric, John E., 1948, Tabulation of copper deposits of California: California Div. Mines Bull. 144, pp. 228-232.

Fairbanks, H. W., 1890, Geology of the Mother Lode region: California Min. Bur. Rept. 10, pp. 23-90.

Fairbanks, H. W., 1894, Geology of a section of El Dorado County: California Min. Bur. Rept. 12, pp. 479-481.

Haley, C. S., 1923, Gold placers of California: California Min. Bur. Bull. 92, 167 pp.

Hamilton, Fletcher, 1922, A review of mining in California during 1921: California Min. Bur. Prelim. Rept. 8, 68 pp.

Hanks, H. G., 1884, Catalogue and description of the minerals of California . . .: California Min. Bur. Rept. 4, pp. 61-397 [El Dorado County limestone, pp. 106-107, 109].

Hinds, N. E. A., 1952, Evolution of the California landscape: California Div. Mines Bull. 158, 240 pp.

Huttl, J. B., 1935a, Montezuma-Apex, newest Mother Lode producer: Eng. and Min. Jour., vol. 136, no. 4, pp. 173-175.

Huttl, J. B., 1935b, Big Canyon's surface plant: Eng. and Min. Jour., vol. 136, no. 5, pp. 216-219.

Irelan, William Jr., 1886, El Dorado County: California Min. Bur. Rept. 6, pt. 2, pp. 42-43.

Irelan, William Jr., 1888, El Dorado County: California Min. Bur. Rept. 8, pp. 164-202.

Irelan, William Jr., 1889, Slate quarrying: California Min. Bur. Rept. 9, pp. 282-283.

Jenkins, O. P., 1932, Geologic map of northern Sierra Nevada: California Div. Mines Rept. 28, pp. 279-298.

Jenkins, O. P., and Wright, W. Q., 1934, California's gold-bearing Tertiary channels: Eng. and Min. Jour., vol. 135, no. 11, pp. 487-502.

Jenkins, O. P., 1935, New technique applicable to the study of placers: California Div. Mines Rept. 31, pp. 143-210.

Jenkins, O. P., 1938, Geologic map of California, scale 1″ = 8 mi., California Div. Mines.

Jenkins, O. P., 1943, Salient geologic events in California and their relationship to mineral deposition: California Div. Mines Bull. 118, pp. 89-93.

Jenkins, O. P., 1948, Geologic history of the Sierran gold belt: California Div. Mines Bull. 141, pp. 23-30.

Knopf, Adolph, 1929, The Mother Lode system of California: U. S. Geol. Survey Prof. Paper 157, 88 pp.

Kunz, G. F., 1892, Mineralogical notes on brookite, octahedrite, and quartz: California Min. Bur. Rept. 11, pp. 207-209.

Lang, Herbert, 1907, The copper belt of California: Eng. and Min. Jour., vol. 84, pp. 909-913, 963-966, 1006-1010.

Lindgren, Waldemar, 1894, U. S. Geol. Survey, Geol. Atlas, Sacramento folio (no. 5), 5 pp., 4 maps.

Lindgren, Waldemar, 1896, U. S. Geol. Survey, Geol. Atlas, Pyramid Peak folio (no. 31), 8 pp., 4 maps.

Lindgren, Waldemar, 1897a, U. S. Geol. Survey, Geol. Atlas, Colfax folio (no. 66), 12 pp., 4 maps.

Lindgren, Waldemar, 1897b, U. S. Geol. Survey, Geol. Atlas, Truckee folio (no. 39), 8 pp., 4 maps.

Lindgren, Waldemar, 1911, The Tertiary gravels of the Sierra Nevada of California: U. S. Geol. Survey Prof. Paper 73, 226 pp.

Lindgren, Waldemar, and Turner, H. W., 1894, U. S. Geol. Survey, Geol. Atlas, Placerville folio (no. 3), 5 pp., 4 maps.

Logan, C. A., 1918, Platinum and allied metals in California: California Min. Bur. Bull. 85, 120 pp. [Upper American River, pp. 29-30].

Logan, C. A., 1921, El Dorado County: California Min. Bur. Rept. 17, pp. 425-433.

Logan, C. A., 1923a, El Dorado County: California Min. Bur. Rept. 18, pp. 44-45 . . . 208-210 . . . 301 . . . 602.

Logan, C. A., 1923b, El Dorado County: California Min. Bur. Rept. 19, pp. 141-143.

Logan, C. A., 1924, El Dorado County: California Min. Bur. Rept. 20, pp. 8-9, 178-179.

Logan, C. A., 1926, El Dorado County: California Min. Bur. Rept. 22, pp. 397-452.

Logan, C. A., 1934, Mother Lode gold belt of California: California Div. Mines Bull. 108, 240 pp. [El Dorado County, pp. 13-54, 207, 209-211].

Logan, C. A., 1935, Review of gold mining in east-central California, 1934: California Div. Mines Rept. 31, pp. 1-23 [El Dorado County, pp. 22-23].

Logan, C. A., 1938, Mineral resources of El Dorado County: California Div. Mines Rept. 34, pp. 206-280, 363-365.

Logan, C. A., 1947, Limestone in California: California Div. Mines Rept. 43, pp. 175-357 [El Dorado County, pp. 222-233].

Murdoch, J., and Webb, R. W., 1948, Minerals of California: California Div. Mines Bull. 136, 402 pp.

Preston, E. B., 1892, El Dorado County: California Min. Bur. Rept. 11, pp. 200-209.

Ransome, F. L., 1897, U. S. Geol. Survey, Geol. Atlas, Mother Lode district folio (no. 63), 11 pp., 8 maps.

Sampson, R. J., and Tucker, W. B., 1931, Feldspar, silica, andalusite, and cyanite deposits of California: California Div. Mines Rept. 27, pp. 407-458 [El Dorado County, pp. 437-438].

Storms, W. H., 1900, The Mother Lode region of California: California Min. Bur. Bull. 18, 154 pp. [El Dorado County, pp. 88-99].

Taliaferro, N. L., 1943, Manganese deposits of the Sierra Nevada, their genesis and metamorphism: California Div. Mines Bull. 125, pp. 277-332.

Taliaferro, N. L., 1951, Geology of the San Francisco Bay counties: California Div. Mines Bull. 154, pp. 117-150 [Geology of Sierra Nevada, pp. 119-120].

Trask, P. D., and others, 1943, Manganese deposits of California . . . : California Div. Mines Bull. 125, pp. 51-215 [El Dorado County, p. 111].

Trask, P. D., and others, 1950, Geologic descriptions of the manganese deposits of California: California Div. Mines Bull. 152, pp. 51-53.

Tucker, W. B., and Waring, Clarence A., 1919, The counties of El Dorado, Placer, Sacramento, Yuba: California Min. Bur. Rept. 15, pp. 267-459 [El Dorado County, pp. 271-308].

[Turner, Mort D.], 1951, El Dorado County: California Div. Mines Rept. 47, pp. 319-321.

Wells, F. G., Page, L. R., and James, H. L., 1940, Chromite deposits of the Pilliken area, El Dorado County, California: U. S. Geol. Survey Bull. 992-O, pp. 417-460.

Whitney, J. D., 1865, El Dorado County: Geological Survey of California, vol. 1, pp. 279-283.

Whitney, J. D., 1880, The auriferous gravels of the Sierra Nevada: John Wilson and Son, Cambridge, Mass., 569 pp.

Young, G. J., 1925a, Mining limestone at Shingle Springs, California: Eng. and Min. Jour. Press, vol. 119, pp. 1001-1002.

Young, G. J., 1925b, Qarrying limestone by blory holes: Eng. and Min. Jour. Press, vol. 120, pp. 13-16.

TABULATION OF EL DORADO COUNTY MINERAL DEPOSITS

The following list of mineral deposits of El Dorado County is arranged alphabetically by commodity and by name of deposit or mine. Numbers which appear in the left-hand column under the heading *Map No.* refer to the Geologic Maps of El Dorado County Showing Mines and Mineral Deposits, plates 10 and 11. The names and numbers in parenthesis in the last column refer to the accompanying bibliography. The first number after the author's name is the year of publication, and is separated from the page reference by a colon. Other references are separated by semicolons. The term "herein" refers to a description in the text.

CHROMITE

MAP NO.	CLAIM, MINE, OR GROUP	OWNER NAME, ADDRESS	SEC.	T.	R.	B & M	REMARKS
	Apex	Not determined	18	13N	11E	MD	One mile southwest of Volcanoville. About 8 tons of ore mined from open pit in 1918. (Cater and others 51:163.)
	Austin		25	13N	10E	MD	See Knoff.
1	Black Oak (Cassiorni)	Not determined	23	12N	10E	MD	Two miles south of Georgetown on ridge west of Traverse Creek. Active 1918 when 36 tons of ore produced and 1942-43 when 107 tons containing 47.5 percent Cr produced. Lenses and pods of coarse chromite in serpentine strike northeast and dip southeast. Developed by open cuts and three 40-ft. shafts. (Bradley and others 18:132; Cater and others 51:161.)
	Bonanza King						Part of Pillikin mine.
2	Bonetti	W. J. Varozza, Shingle Springs	7	8N	10E	MD	Three and one-half miles east of Latrobe and northwest of Big Canyon Creek. Active 1917 and 1942. A 1- to 3-ft. lens of chromite 60 ft. long in dunite strikes N. 20E. and dips 70° SE; estimated to contain 200 tons of ore. (Bradley and others 18:132; Cater and others 51:123.)
	Brandon	Robert D. Domecq et al., Route 1, Box 39, Shingle Springs	12	8N	9E	MD	Three miles east of Latrobe on ridge between Hungry Hollow and Indian Creeks. Active 1918 when four carloads of ore produced. A series of northwest-trending chromite pods in dunite. Developed by open cuts. (Bradley and others 51:122-123.)

CHROMITE (CONT.)

MAP NO.	CLAIM, MINE, OR GROUP	OWNER NAME, ADDRESS	SEC.	LOCATION T.	R.	B & M	REMARKS
3	Bryant	Stanley S. Bryant, R.F.D. Box 38, Shingle Springs	13	8N	9E	MD	On west bank of Big Canyon Creek $2\frac{1}{2}$ miles south of Brandon Corner. Active 1918. A northeast-trending ore body. (Cater and others 51:122.)
	Burnett	F. L. Burnett, 469 S. Washington, Placerville	3 34	10N 11N	8E 8E	MD MD	One mile southwest of Salmon Falls on the north side American River. Active in 1918 when 139 tons of ore produced. Layered and disseminated bodies of chromite that strike northwest occur in a zone several hundred feet long. Developed by open cuts and shallow shafts. (Allen 41:138; Cater and others 51:151-152.)
	Buzzard Mill	A. C. Darrington, Repressa					Located on Darrington Ranch. During World War 1, chromite concentrated in 5-stamp mill with Overstrom concentrator. (Bradley and others 18:132.)
	Cassiorni						See Black Oak.
4	Central Pacific Railroad	D. O. Niegel, Cool	23	12N	9E	MD	Two miles southwest of Greenwood. Active 1916-18 when about 250 tons of ore were mined. A northwest-trending lens of 35-41% ore occurs in peridotite. Developed by open cuts. (Cater and others 51:160.)
	Central Railroad	United States of America (withdrawn for federal reservoir site)	21	11N	8E	MD	One mile northwest of Pillikin mine. A northwest-striking zone of low-grade disseminated ore developed by adits and 200-ft. shaft. (Allen 41:38.)

CHROMITE (CONT.)

MAP NO.	CLAIM, MINE, OR GROUP	OWNER NAME, ADDRESS	SEC.	T.	R.	B & M	REMARKS
5	Chaix	Roy Chaix, et al., Box 87, Placerville	14	8N	9E	MD	(Bradley and others 18:132-133; Averill 43:140; Cater and others 51:124; herein.)
	Chrome Divide	Georgetown Lumber Co., Georgetown	25	13N	10E	MD	Three miles northeast of Georgetown. Active 1941-43 when 51 tons of chromite produced Pods and lenses of chromite in serpentine strike northwest. (Cater and others 51:162.)
	Chrome Gulch						Part of Pillikin mine.
	Cowell	E. F. Glenn, et al., Folsom	4?	9N	9E	MD	Three miles east of Clarksville. Active during World War I when four carloads of chromite were produced. (Bradley and others 18:133; Cater and others 51:125-126.)
6	Darrington (Gurney)	George Darrington, Folsom	33	11N	8E	MD	(Allen 41:38; Averill 43:317-318; 556-557; Cater and others 51:150-151; herein.)
	Dickson	E. F. Glenn, et al., Folsom	4	9N	9E	MD	Three miles east of Clarksville. Northwest-trending chromite pods in dunite were mined in an open cut. (Cater and others 51:125.)
7	Dobbas	D. J. Dobbas, Auburn	22	11N	8E	MD	(Allen 41:39; Cater and others 51:146-148; herein.)
	Donnelly		21	11N	8E	MD	Part of Pilliken mine. (Bradley and others 18:133.)

CHROMITE (CONT.)

MAP NO.	CLAIM, MINE, OR GROUP	OWNER NAME, ADDRESS	SEC.	T.	R.	B & M	REMARKS
	Ever	Not determined	32	9N	9E	MD	Near Cothrin Station. Prospected in 1918. Small streaks and lenses of chromite strike north, occur in 100-ft. zone. (Cater and others 51:125.)
	Expansion						See Swortfiguer.
	Forni	Not determined		8N	10E	MD	Four miles east of Latrobe. One ton of ore produced in 1918. (Bradley and others 18:133; Cater and others 51:125.)
	Freeman	I. Ellingham and Erwin W. Vogel, c/o County Courthouse, Placerville	13	8N	9E	MD	Four miles southeast of Latrobe. In 1918, 40 tons of ore produced from north-striking chromite lens in serpentine. Developed by open cut. (Bradley and others 18:133-134; Cater and others 51:122.)
	Glenn	Willard C. Egloff, et al., Box 506, Folsom	14	8N	9E	MD	Adjoining Murphy mine. Prospected in 1918. (Bradley and others 18:134.)
	Gold Bug	Theresa Garibaldi, Amador City	19	8N	10E	MD	Four miles southeast of Latrobe on ridge east of Big Canyon Creek. Active in 1918 when 16 tons of chromite produced. A northwest-striking, 1- to 3-ft. zone of low-grade ore, 125 feet long. Developed by open cuts. (Cater and others 51:121.)

CHROMITE (CONT.)

MAP NO.	CLAIM, MINE, OR GROUP	OWNER NAME, ADDRESS	LOCATION				REMARKS
			SEC.	T.	R.	B & M	
	Gordon	N.P. and Camilia Runge Route 2, Box 251, Placerville	13	10N	9E	MD	Four miles north of Shingle Springs. In 1918, 31 tons of chromite produced from float. (Cater and others 51:153.)
	Granite Bar						See Hoosier Gulch Placers. (Cater and others 51:148.)
8	Green	Not determined	19	13N	11E	MD	One and one-half miles south of Volcanoville near Otter Creek. Worked originally for gold. Active 1917-18 when more than 110 tons chromite produced and 1942 when 64 tons produced. A chromite lens up to 7 feet wide developed by 350-ft. adit, 40-ft. shaft, and raises and crosscuts. (Bradley and others 18:134-135; Cater and others 51:162-163.)
	Green claim	Not determined	24	12N	10E	MD	Two miles southeast of Georgetown. Active 1918 when 17 tons of 51% Cr$_2$O$_3$ ore produced. Pods strike northwest. Developed by 15 ft. shaft and drift. (Cater and others 51:161.)
	Gurney						See Darrington.
	Hector Williamson						See Williamson.
9	Helemar	Theresa Garibaldi, Amador City	30	8N	10E	MD	Five miles southwest of Latrobe. Active 1944-45 when 57 tons of ore containing 38% Cr$_2$O$_3$ produced. (Cater and others 51:121.)

CHROMITE (CONT.)

MAP NO.	CLAIM, MINE, OR GROUP	OWNER NAME, ADDRESS	LOCATION				REMARKS
			SEC.	T.	R.	B & M	
	Henser	Not determined	14	12N	10E	MD	Two miles southeast of Georgetown. In 1918, 13 tons of chromite produced. (Cater and others 51:161.)
	Hill-Top Chrome	Not determined	18	13N	11E	MD	One mile southwest of Volcanoville. In 1918, 7 tons ore produced. Developed by 22-ft. shaft. (Cater and others 51:163.)
10	Hoff	Theresa Garibaldi, Amador City	30	8N	10E	MD	Five miles southeast of Latrobe. Active World War I when 107 tons 28 to 33½ Cr$_2$O$_3$ ore produced. A. northwest-striking chromite lens 60 ft. long in sheared serpentine developed by open cut. (Bradley and others 18:135-136; Cater and others 51:120-121.)
	Hoosier Gulch Placers (Granite Bar)	Hoosier Gulch Placers, Sacramento	20	11N	8E	MD	Chromite-bearing river gravel at Granite Bar on American River near Pillikin mine mined in 1942 with dragline. (Cater and others 51:148.)
	Irish	Not determined	20	10N	10E	MD	Two and one-half miles east of Rescue. In 1918, 18 tons ore produced. Small chromite pods and stringers developed by open cuts. (Bradley and others 18:136; Cater and others 51:153.)
11	Joerger	Bertha J. Burton, 1115 Yale, Fresno	35	10N	8E	MD	(Bradley and others 18:136; Averill 43:140; Cater and others 51:126; herein.)

CHROMITE (CONT.)

MAP NO.	CLAIM, MINE, OR GROUP	OWNER NAME, ADDRESS	SEC.	T.	R.	B & M	REMARKS
12	Kelly	United States of America (withdrawn for Federal reservoir site).	16	11N	8E	MD	Just east of Rattlesnake Bridge. Active in 1918 when 25 tons 28% ore produced. (Bradley and others 18:136.)
	Knoff (Austin)	W. B. and W. E. LaDue, 2319 "O" Street, Sacramento	25	13N	10E		Three miles northeast of Georgetown on Little Bald Mountain. Active 1918 when 400 tons of ore mined, and 1942-44 when 79 tons mined. Pods and lenses of chromite in sheared serpentine strike northwest. Developed by open cuts and shallow shafts. (Bradley and others 18:131; Cater and others 51:162.)
	Laicey	Not determined	32	12N	10E	MD	One and three-quarters miles west of Garden Valley. Small chromite pods in serpentine and talc. Developed by open cuts. (Cater and others 51:159.)
	McCurdy	Rolland Armstrong, 1291 Crow Canyon Road, Hayward	8	11N	10E	MD	Two miles north of Coloma. Active 1918 when 200 tons of ore containing 36% Cr₂O₃ produced. Two groups of north-east-striking chromite lenses in serpentine and talc. Developed by open cuts and 45-ft. shaft. (Bradley and others 18:136; Cater and others 51:155-156.)
13	McDonald and Buys	Nick J. Schubin, et al., Route 2, Box 216, Placerville	6	10N	10E	MD	One mile south of Four Corners. Active 1918 when 350 tons of chromite produced. Chromite pods strike north. Developed by three shafts and open pit. (Cater and others 51:54.)
	Miller	Joseph H. Miller, P.O. Box 282, Folsom	35	10N	8E	MD	One and one-half miles northwest of Clarksville, near Walker mine. Small amounts of layered chromite. (Cater and others 51:128.)

CHROMITE (CONT.)

MAP NO	CLAIM, MINE, OR GROUP	OWNER NAME, ADDRESS	SEC.	T.	R.	B & M	REMARKS
15	Murphy	Willard C. Egloff, Box 506, Folsom	14	8N	9E	MD	(Bradley and others 18:136-137; Cater and others 51:123-124; herein.)
	Nielson						Part of Pillikin mine.
	Noble Electric Steel Company						Operated chrome mill one mile southwest of Salmon Falls during World War I. Ore was crushed, fine-ground, and concentrated on Wilfley tables and Senn concentrator. (Bradley and others 18:137.)
16	O'Brien (S-Bend)	C. F. Gladwill, 381 Acacia Street, Vacaville	6	11N	10E	MD	Two miles north of Coloma west of Perry Creek. Active 1918 when several hundred tons of ore produced, and 1942 when 3,000 tons produced and treated at Volo Mill at Shingle Springs. Lens of chromite up to 20 feet wide in dunite. Developed by two adits and glory hole. (Cater and others 51:126.)
	Ogle	Not determined	18	13N	11E	MD	One mile south of Volcanoville. Active 1917 when 47 tons of 45 percent ore produced. North-striking chromite lenses in serpentine. Developed by open cut. (Bradley and others 18:137; Cater and others 51:163, 166.)
17	Pfeiffer	W. J. Varozza, Shingle Springs	7	8N	10E	MD	Three miles east of Latrobe west of Big Canyon Creek. Active 1918 when 80 tons of chromite produced; prospected in 1942. Chromite pods trend northeast. Developed by open cuts. (Bradley and others 18:140; Carter and others 51:22.)
	Pilliken						See Pillikin.

CHROMITE (CONT.)

MAP NO.	CLAIM, MINE, OR GROUP	OWNER NAME, ADDRESS	SEC.	T.	R.	B & M	REMARKS
18	Pillikin (Pilliken. Also includes Bonanza King, Chrome Gulch, Donnelly, Nielson, and Steele.)	American Trust Co., 464 California Street, San Francisco	21, 28	11N	8E	MD	(Lindgren 94:3; Tucker 19:274,275; Bradley and others 18:137-140, 141-143; Logan 20:431-432; 26:404; 38: 207,212; Allen 41:136-138; Averill 43:72; Cater and others 51:131-146; herein.)
19	Pilot Hill	W. J. Rogers, Route 3, Box 1340, Modesto	12	11N	8E	MD	Just west of Pilot Hill summit. Active in 1916 when 200 tons ore produced. Lenticular ore body in dunite strikes northeast. Developed by open cut. (Bradley and others 18:140-141; Allen 41:135-136; Cater and others 51:154-155.)
	Placer Chrome Company						Worked Pillikin mine during World War I; see Pillikin mine herein.
	Ruby Consolidated	Not determined	17,18 19	13N	11E	MD	One mile south of Volcanoville. A gold and chrome mine. During World War I, three carloads chromite containing 46 percent Cr_2O_3 shipped. See also gold tabulated list.
	S-Bend						See O'Brien.
20	Shelly (Wolf)	J. A. Wolf, Garden Valley	5	11N	10E	MD	Two miles southwest of Garden Valley. Active 1918 when 1,284 tons of ore containing 30 percent Cr_2O_3 were mined. Irregular lenses of chromite in serpentinized dunite trend north. Developed by open pit and shaft. (Bradley and others 18:143; Cater and others 51:158.)

CHROMITE (CONT.)

MAP NO.	CLAIM, MINE, OR GROUP	OWNER NAME, ADDRESS	SEC.	T.	R.	B & M	REMARKS
	Sheppard	Not determined	32	12N	10E	MD	Two miles west of Garden Valley. Active World War I when more than 50 tons of 35 percent Cr₂O₃ ore produced. North-trending chromite lenses in serpentine. Developed by open cuts and shafts. (Cater and others 51:159-160.)
	Simon	C. F. Gladwill, 381 Acacia Street, Vacaville	5	11N	10E	MD	Two miles southwest of Garden Valley. Active 1918, 1920 94 tons 35 percent Cr₂O₃ ore produced. Northeast-trending chromite lenses in serpentinized dunite. Developed by open cuts. (Cater and others 51:156-158.)
21	Simpson	Not determined	13	10N	8E	MD	Five miles north of Clarksville. Active in 1917 when 54 tons ore produced. Lenticular chromite bodies striking north near serpentine-schist contact. Developed by open cuts and shallow shafts. (Bradley and others 18:143; Cater and others 51:128.)
	Smith	George A. Smith et al., Box 66, Folsom	9	8N	9E	MD	Three-quarters of a mile northwest of Latrobe. North-west-striking chromite lens in serpentine and talc mined from 100-ft. open cut. (Cater and others 51:124-125.)
	Southeastern Railroad	United States of America (withdrawn for Federal reservoir site.)	21	11N	8E	MD	One mile north of Pillikin mine. Zones of small chromite pods developed by open pits and adits. (Allen 41:138.)
22	Stafford	Ethel S. Hughes et al., R.F.D. #1, Box 568, Carmel	36	13N	10E	MD	Two miles northeast of Georgetown. Active 1918 when 198 tons ore produced and 1942-43 when a few tons produced. Irregular north-striking chromite lenses and pods mined in open cuts. (Cater and others 51:161-162.)
	Steele		28	11N	8E	MD	See Pillikin mine.

CHROMITE (CONT.)

MAP NO.	CLAIM, MINE, OR GROUP	OWNER NAME, ADDRESS	LOCATION				REMARKS
			SEC.	T.	R.	B & M	
	Stifle	Not determined	25	12N	10E	MD	Three miles northwest of Garden Valley. Active 1918 when 4 tons 35 percent ore produced. Small chromite pods in serpentine and talc developed by open cuts. (Bradley and others 18:143; Cater and others 51:160-161.)
	Swortfiguer (Expansion)	George B. and Helen Swortfiguer, 804-48th Street, Sacramento	17,18	10N	10E	MD	Four miles northeast of Shingle Springs. Active 1914-18 when 80 tons ore produced; prospected in 1943. Developed by open cut. (Averill 43:72.)
	Taylor						See Darrington.
	Thomas and Meldrum	Not determined	18	10N	10E	MD	Two miles east of Rescue. During World War I, one carload ore produced from small stringers and pods in serpentine. Developed by two shafts. (Cater and others 51:153.)
	Trio Chrome Company						During World War I, leased Hoff and Helemar properties and placer mined chromite-bearing gravel near Cosumnes River in sec. 25, T. 8 N., R. 9 E. (Cater and others 51:121.)
	Tropper	Not determined	32	12N	10E	MD	One and one-half miles west of Garden Valley. Active 1918 when 110 tons ore mined. Lens 10-ft. long developed by 40-ft. inclined shaft. (Bradley and others 18:143-144.)

CHROMITE (CONT.)

MAP NO.	CLAIM, MINE, OR GROUP	OWNER NAME, ADDRESS	SEC.	T.	R.	B & M	REMARKS
	Veerkamp	Ray A. Veerkamp, Garden Valley	33	12N	10E	MD	One and one-half miles southwest of Garden Valley. Active 1916 when 38 tons ore containing 41 percent Cr2O3 mined. Number of small northwest-striking chromite pods developed by open pits. (Cater and others 51: 158-159.)
23	Walker	Faustino Silva, Route 8, Box 951, Sacramento	35	10N	8E	MD	(Cater and others 51:126-128; herein.)
	Wiley	Not determined	1	10N	9E	MD	One mile southwest of Four Corners. Active 1916 when 45 tons chromite produced from open cuts. (Bradley and others 18:144; Cater and others 51:154.)
24	Williamson	George Williamson et al., Route 2, Box 288, Placerville	7	10N	10E	MD	Six miles due north of Shingle Springs. Active 1918 when 55 tons ore containing 40 percent Cr2O3 produced. Irregular chromite pods in slickentite. (Cater and others 51:154.)
	Williamson (Hector Williamson)	Not determined	7	10N	10E	MD	Possibly same as above, though production of only 40 tons reported. (Cater and others 51:153.)
	Wilson	Not determined	23	12N	10E	MD	Two miles northwest of Garden Valley. Shipping and milling-grade ore produced 1943. (Averill 43:72.)
	Wolf						See Shelly.
	Zanini	Not determined	35	9N	9E	MD	Two miles northeast of Latrobe. Low-grade deposit prospected during World War I. (Bradley and others 18: 144.)

COPPER

MAP NO.	CLAIM, MINE, OR GROUP	OWNER NAME, ADDRESS	LOCATION					REMARKS
			SEC.	T.	R.	B & M		
	Agara	Clifford Smith, Somerset	24	9N	12E	MD		Three miles northeast of Fairplay and just north of Consumnes copper mine. Developed by 25-ft. shaft. (Aubury 02:180; 08:216; Eric 48:228.)
25	Alabaster Cave	United States of America (withdrawn for Federal reservoir site)	10,15	11N	8E	MD		One mile east of Rattlesnake Bridge. Active prior to 1902. A north-striking 8-ft. copper-bearing zone dips east with limestone hanging and slate footwall. Contains 3 to 4 percent copper and some gold and silver. Developed by 300- and two 50-ft. shafts and 100- and 30-ft. adits. (Aubury 05:176; Heyl and others 48:228; Aubury 08:211-212; Tucker 19:276.)
	Arizona	Not determined	24	12N	10E	MD		Two miles southeast of Georgetown. Copper claim containing gossan outcrops as much as 100 ft. wide. (Aubury 05:180; 08:216; Eric 48:228.)
	Barklage and Miller	William Barklage et al., 5713 State Avenue, Sacramento	13	12N	10E	MD		Two miles southeast of Georgetown. Active 1908. Gossan capping 100 ft. wide in slate. Developed by 118-ft. adit. (Aubury 05:178; 08:214; Eric 48:228.)
26	Big Buzzard (Hercules, Darrington)	George Darrington, Folsom	29	11N	8E	MD		(Logan 21:430; 23a:209; 23b:141,142; 26:406,412; Averill 43:72; herein.)
	Blue Cat (Madelina, Madeline, Magdalena)	George Fausel, Placerville	18	9N	11E	MD		Five miles south of Diamond Springs; south extension of Noonday mine. Active prior to 1900. Zone 40 to 60 ft. wide in metasediments contains pyrrhotite, chalcopyrite, and pyrite with gold. Developed by 90-ft. cross-

COPPER (CONT.)

MAP NO.	CLAIM, MINE, OR GROUP	OWNER NAME, ADDRESS	SEC.	T.	R.	B & M	REMARKS
	Blue Cat (continued)						cut adit, 100-ft. crosscut adit, 100-ft. drift, and 105-ft. shaft. (Storms 00:91; Eric 48:228.)
	Bob						See Iron Crown.
27	Boston	Not determined	22	9N	9E	MD	Four miles southwest of Shingle Springs. Active 1860's and 1870's when good ore was produced. Country rock schist. Developed by 400-ft. shaft. (Aubury 05:180; 08:216; Eric 48:229.)
28	Breala	Not determined	2	8N	9E	MD	Two miles northeast of Latrobe. Old shaft reopened in 1917 and some 4½ percent Cu ore mined. Developed by 70-ft. shaft. (Logan 26:407; Eric 48:229.)
	Bryant Ranch	Not determined	2	8N	9E	MD	One and one-half miles northeast of Latrobe. Active 1860. A 4-ft. vein containing copper oxide and carbonates. Developed by 64-ft. shaft. (Aubury 05: 180; 08:216-217; Eric 48:229.)
	Bunker Hill	E. A. Long et al., Renobscot Farm, Cool	14	12N	9E	MD	Two miles southwest of Greenwood. Active 1860's. Developed by 60-ft. shaft. (Aubury 05:181.)
29	Cambrian	L. D. Stodick, et al., Lotus	23	11N	9E	MD	Three miles west of Coloma. Active 1850's, 1900, 1908. Three veins of talcose schist between granodiorite and serpentine contain chalcopyrite, malachite, native copper and gold. Ore contains up to 10 percent

COPPER (CONT.)

MAP NO.	CLAIM, MINE, OR GROUP	OWNER NAME, ADDRESS	SEC.	T.	R.	B & M	REMARKS
							copper. Developed by 113-, 220-, and 1360-ft. adits, winzes and drifts. (Aubury 05:177-178; 08:213-214, 218; Tucker 19:276-277; Eric 48:229.)
30	Camelback (Voss)	Aleta V. Coulet, 215 Market Street, San Francisco	11	11N	8E	MD	Three miles southwest of Pilot Hill on Burner Hill. Active around 1920. A massive quartz vein strikes northeast, dips northwest, and contains chalcopyrite, pyrite, and some bornite. Developed by 25-ft. shaft, 123-ft. drift adit, 165-ft. crosscut adit. A parallel vein half a mile east developed by 200-ft. and 40-ft. shafts. (Logan 20:430-431; 26:407; Eric 48:229.)
31	Contraband (Ford)	Not determined	24	12N	10E	MD	Two miles southeast of Georgetown. Active 1860, 1902, 1910. A 12-ft. vein striking north 80° east and dipping 45° northwest in schist contains native copper and copper sulfides. Also an asbestos property (see section on asbestos). (Aubury 05:178; 08:214-216; Tucker 19:277; Eric 28:229.)
	Copper Chief	Not determined	12	12N	10E	MD	Two miles east of Georgetown. Gossan outcrops 100 to 200 ft. wide. (Aubury 05:180; 08:216; Eric 48:229.)
	Costa Ranch	F. J.Costa, Jr., Pilot Hill	12	11N	8E	MD	Two miles southwest of Pilot Hill. North-striking east-dipping vein contains malachite, chalcopyrite, and pyrite. Developed by 60-ft. vertical shaft and open cuts. (Aubury 08:218; Eric 48:229.)
32	Cosumnes	Clifford Smith, Somerset	24,25	9N	12E	MD	(Crawford 96:58; Aubury 05:178; 08:214,215; Tucker 19:277; Eric 48:229; herein.)

COPPER (CONT.)

MAP NO.	CLAIM, MINE, OR GROUP	OWNER NAME, ADDRESS	LOCATION				REMARKS
			SEC.	T.	R.	B & M	
	Cothrin	Not determined	29	9N	9E	MD	Near Cothrin Station. Prospected 1926. Massive pyrite with chalcopyrite in amphibolite. Developed by 100-ft. shaft. (Logan 26:407; Eric 48:229.)
	Darrington						See Big Buzzard.
	Diamond Springs						See Larkin in Lode gold section.
	Dr. Wren	C. Padilla, El Dorado	7	9N	11E	MD	Three miles southeast of El Dorado and east of the Mother Lode. A 6-ft. vein of talcose schist contains 5 to 18 percent copper. Developed by 18-ft. shaft. (Aubury 05:180; 08:216; Eric 48:229.)
	Dodson						See Rip and Tear.
33	E. E. Copper	Not determined	18	9N	11E	MD	Four miles southeast of El Dorado. Active around 1914. A 2-ft. north-striking and east-dipping vein in metadiabase contains bornite, chalcopyrite, pyrite, gold, and silver. Developed by 85-ft. vertical shaft, with 300 ft. of drifts and 100- and 300-ft. adits. (Aubury 08:218-219; Tucker 19:277; Eric 48:230.)
34	El Dorado (Roosevelt)	Calivada Development Company, Box 4, Garden Valley	34	12N	10E	MD	(Bedford 46; Eric 48:230; herein.)
	Ford						See Contraband.

COPPER (CONT.)

MAP NO.	CLAIM, MINE, OR GROUP	OWNER NAME, ADDRESS	SEC.	T.	R.	B & M	REMARKS
	Funny Bug	R. V. Montgomery, Route 2, Box 193-K, Placerville	4	10N	10E	MD	Gold mine containing some copper; see Lode gold section. (Logan 38:242-243; Eric 48:230.)
	Hale	Clifford Smith et al., Somerset	25	9N	12E	MD	South extension of Cosumnes mine. (Aubury 08:217; Eric 48:230.)
	Hercules						See Big Buzzard.
	Homestead	Not determined		12N	9E	MD	Three miles west of Greenwood. A 14-ft. vein contains copper and gold. (Aubury 05:177; 08:213; Eric 48:230.)
	Irland	Not determined	15	10N	10E	MD	Three miles west of Placerville. Active 1866, 1906. Vein in granodiorite strikes north, dips east, contains chalcopyrite. Ore contained 2 percent copper, some gold and silver. Developed by 75-ft. vertical shaft and 18-ft. drift. (Aubury 08:218; Eric 48:230.)
35	Iron Crown (Bob)	Not determined	13	12N	10E	MD	One mile southeast of Georgetown. Active prior to 1902 and again around 1908. A series of gossan cappings and copper-bearing veins with slate and serpentine walls; extends south to Contraband mine. Water is copper-bearing. Developed by 75-ft. shaft and open cuts. (Aubury 05:181-182; 09:219-220; Tucker 19:276; Eric 48:230.)

COPPER (CONT.)

MAP NO.	CLAIM, MINE, OR GROUP	OWNER NAME, ADDRESS	LOCATION SEC.	T.	R.	B & M	REMARKS
	Larkin (Diamond Springs)		29	10N	11E	MD	Gold mine containing some copper; see section on Lode gold.
	Lilyama (Little Emma, Volo)						See Pioneer-Lilyama.
	Little Emma						See Pioneer-Lilyama. (Aubury 08:212; Logan 26:408.)
	Longshot	Leslie Fry and Clifford Smith, Somerset	26	9N	12E	MD	One mile west of Cosumnes copper mine. Prospect that is worked intermittently. Copper stains in granite and metasediments. Developed by 200-ft. adit.
	Madelina						See Blue Cat.
	Madeline						See Blue Cat.
	Magdalena						See Blue Cat.
36	Noonday	George Fausel, Placerville	18	9N	11E	MD	(Aubury 02:182; 08:220; Tucker 19:278; Eric 48:231; herein.)
37	Pioneer-Lilyama (Little Emma, Volo)	H.H. Mitchel 9490 Brighton Way, Beverly Hills; Leased by Wilcox-Lilyama Mining Co., F.L. Wicks, Manager	3	11N	9E	MD	(Aubury 02:176-177; 08:212-213, 218; Tucker 19:277-278; Averill 43:140; Cox and others 48:43-47; Eric 48:231; herein.)
	Revoir	Not determined	12	9N	12E	MD	Two miles southeast of Pilot Hill. Copper prospect south of Costa Ranch mine. (Aubury 08:217; Eric 48:231.)

COPPER (CONT.)

MAP NO.	CLAIM, MINE, OR GROUP	OWNER NAME, ADDRESS	LOCATION			B & M	M	REMARKS
			SEC.	T.	R.			
38	Rip and Tear (Dodson)	Fred H. Dodson, 2404 - 26th Street, Sacramento	3	8N	9E	MD		(Aubury 02:181; 08:219; Logan 26:408; Averill 43:140; Eric 48:231; herein.)
39	Robert	Robert L. Cameron et al., 1217 Del Paso Blvd., North Sacramento	13	9N	11E	MD		Two miles northeast of Outingdale. A 3½-ft. vein in schist and slate; ore yielded 4 to 24 percent copper. Developed by 80-ft. shaft and 150-ft. crosscut adit. (Aubury 05:180-181; 08:216-217; Tucker 19:278; Eric 48:232.)
	Roosevelt							See El Dorado.
	Seven Bells (Sporting Boy)	Not determined	10	10N	10E	MD		Four miles west of Placerville. Prospected 1917 and 1918. Vein up to 18 inches wide contains copper and gold. Developed by 65-ft. shaft. (Logan 26:408; Eric 48:232.)
	Sporting Boy							See Seven Bells.
	Volo							See Pioneer-Lilyama and Volo mill.
	Volo Mill	Volo Mining Company, 464 Main Street, Placerville	21	10N	10E	MD		Also see Shaw mine in Lode gold section. (Averill 43:140; herein.)
	Voss							See Camelback.
	Woods	George A. Smith et al., Box 66 Folsom	4	8N	9E	MD		One mile northwest of Latrobe. A 5-ft. vein in schist contains copper. Developed by 2-ft. shaft. (Aubury 05:181; 08:219; Eric 48:232.)

LODE GOLD

MAP NO.	CLAIM, MINE, OR GROUP	OWNER NAME, ADDRESS	SEC.	T.	R.	B & M	REMARKS
40	Adams Gulch (Stony Point, Sullivan)	Camilla D. Head, 624 South 13th St., San Jose	25,26	9N	10E	MD	On Mother Lode two miles northeast of Nashville. Active 1902-11, when $9,482 was produced, and again in 1914. A 4-ft. vein in slate strikes north 10° west and 60° east. Developed by 180- and 200-ft. crosscut adits. (Crawford 94:101; 96:132; Tucker 19:279; Logan 34:15.)
41	Adjuster (Hustler)	Cecilia Simpson, 4029 Kuhrle Street, Oakland	12	9N	10E	MD	On Mother Lode two miles southwest of El Dorado. Active prior to 1914. A 5-ft. vein in slate. Developed by 250-ft. crosscut adit and about 100 feet of drifts. A 10-stamp mill on property. (Tucker 19:280; Logan 34:15.)
42	Alhambra	Wilber E. Timm et al., c/o County Court House, Placerville Leased by Alhambra-Shumway Mines Inc., 681 Market Street, San Francisco Sub-leased to Alhambra Gold Mine Corporation, 1930 Outpost Drive, Hollywood	6,7	11N	11E	MD	(DeGroot 90:178; Logan 35:22; 38:216; Bowen and Cripden 48:70; herein.)
43	Alpine (Union Consolidated)	Elizabeth F. Burks, 4240 12th Street, Sacramento	15,16	12N	10E	MD	(Irelan 88:167-168; Crawford 96:132; Tucker 19:280; Logan 34:15-17; Bowen and Crippen 48:70; herein.)

LODE GOLD (CONT.)

MAP NO.	CLAIM, MINE, OR GROUP	OWNER NAME, ADDRESS	SEC.	T.	R.	B & M	REMARKS
	Artic	Not determined	20	9N	13E	MD	Two miles southwest of Grizzly Flat. Long idle.
	Argonaut (Aultman, Golden Unit)	El Dorado Argonaut Mines, c/o J. A. Sisler, Box 527, Visalia	17	12N	10E	MD	On **Mother** Lode 1½ miles southeast of Greenwood. Active 1880's, 1921, and 1927-28. A northwest striking vein up to 15 ft. wide developed by drift adit. Ore yielded up to $15 per ton. Ore first treated in 10-stamp mill and later in a Gibson mill. (Irelan 88:176; DeGroot 90:176; Crawford 94:101; 96:132; Tucker 19:280; Hamilton 22:29; Logan 22:209; 26:413-414; 34:17.)
44	Argonaut Fraction	Pete Lopez, 4833 - 9th Avenue, Sacramento	17	12N	10E	MD	(Herein.)
	Armstrong	Not determined	28	9N	13E	MD	Two and one-half miles south of Grizzly Flat. Prospected in 1920. North-and east-striking veins in granodiorite. Developed by 100- and 114-ft. shafts and 180-ft. drift. (Logan 20:425-426.)
	Asbestos						See Brust.
	Aultman						See Argonaut.
	Bald Eagle						See Crown Point Consolidated.
	Baldwin						See Briarcliffe. (Crawford 94:102, 96:133; Tucker 19:280.)

LODE GOLD (CONT.)

MAP NO.	CLAIM, MINE, OR GROUP	OWNER NAME, ADDRESS	LOCATION SEC.	LOCATION T.	LOCATION R.	LOCATION B & M	REMARKS
45	Balmaceda	Camilia D. Heald, 624 South 13th Street, San Jose	35	9N	10E	MD	One and one-half miles northeast of Nashville. Active in 1914. Two parallel 4-ft. veins in Mariposa slate strike north 20° east and dip 65° east; ore shoots were 40 to 100 feet long. Developed by 500-ft. drift adit on west vein; stoped to surface. (Tucker 19:280-281.)
46	Baltic	Oscar E. Reeg, 12 Mill Street, Placerville	23	10N	13E	MD	Five miles north of Grizzly Flat on north side of Baltic Peak. Active in 1896 and 1907. A 1-ft. vein in slate strikes northeast and dips 50° East. Developed by 500-ft. drift adit and 130-ft. inclined shaft. Ore treated in 10-stamp mill. (Crawford 96:133; Tucker 19:280.)
47	Barnes-Eureka (Greenstone)	B. F. Baskin, Sr., Route 2, Box 38, Placerville	28 33	10N	10E	MD	(Crawford 94:102,112; 96:144; Tucker 19:281; Logan 38:217; herein.)
	Base Bonanza	Not determined	32	12N	10E	MD	One mile west of Garden Valley. Active prior to 1894. Diorite is east and serpentine is west of vein. (Crawford 94:102.)
	Bathurst						See Coe Hill.
48	Beebe (Brooklyn, East Lode, Iowa, Woodside-Eureka)	Woodside-Eureka Mining Co., Ltd., c/o T. B. Meyers, 1605 Tribune Tower, Oakland 12	2,3, 11	12N	10E	MD	(Logan 34:17-19, 207-211; 38:217-218; Bowen and Crippen 48:70; herein.)

LODE GOLD (CONT.)

MAP NO.	CLAIM, MINE, OR GROUP	OWNER NAME, ADDRESS	LOCATION SEC.	T.	R.	B & M	REMARKS
	Bell						See River Hill group.
	Berg						See Blue Gouge.
	Berry						See Placerville Gold Mining Company.
49	Bidstrup	Walter Bidstrup, Box 35, El Dorado	11	9N	10E	MD	Two miles south of El Dorado. A 1-ft. north-striking vein in granodiorite. Developed by 35-ft. shaft and 100-ft. adit. (Crawford 94:103; 96:133; Tucker 19:281.)
	Big Buzzard						Copper mine containing some gold; see copper section.
50	Big Canyon (Oro Fino)	Capital Company, 1 Powell Street, San Francisco 2	29	9N	10E	MD	(Irelan 88:174-175; Crawford 94:103-104; 96:133; Tucker 19:293; Logan 22:209; 26:412; Huttl 35:216-219; Logan 35:22; 38:220-223; herein.)
51	Big Chunk	T. C. Smith, Jr., 100 Canal Street, Placerville	24	11N	10E	MD	Half mile east of Kelsey. A 3-ft. vein in hanging wall of Mother Lode. Developed by 100-ft. shaft and 150-ft. adit. (Tucker 19:281; Logan 34:19.)
	Big Four (Golden Oak)	Not determined	34	12N	10E	MD	On Mother Lode one mile southeast of Garden Valley. Active during 1890's; prospected in 1940. A 30-inch vein in slate strikes north 10 west and dips 55° east. Ore yielded $10 to $13 per ton. Developed by 96-ft. inclined shaft and 100-ft. drift. (Tucker 19:281; Logan 34:19.)

LODE GOLD (CONT.)

MAP NO.	CLAIM, MINE, OR GROUP	OWNER NAME, ADDRESS	SEC.	T.	R.	B & M	REMARKS
	Big Jim (Phillips)	Not determined		8N	8E	MD	Two and one-half miles southwest of Latrobe. Active around 1896. A north-trending vein with gouge lies between slate and diabase. Developed by shaft and 240-ft. crosscut adit. Ore treated in 2-stamp mill. (Crawford 96:134-135.)
52	Big Sandy (James Marshall)	H. J. Picchetti, 1115 Fairview Avenue, San Jose	24	11N	10E	MD	(DeGroot 90:173-174; Crawford 94:104; 96:134; Tucker 19:281; Logan 34:19-20; herein.)
	Black Hawk	H. J. Picchetti, 1115 Fairview Avenue, San Jose	24	11N	10E	MD	On Mother Lode 3/4-mile south of Kelsey. A 4-ft. vein in slate. Developed by 200-ft. drift adit. (Tucker 19:281.)
53	Black Oak (Clark, Davey, Dayton Consolidated)	Russell J. Wilson Garden Valley	27,34	12N	10E	MD	(Logan 34:20-21; 38:223-224; Bowen and Crippen 48:70; herein.)
	Black Lead	Not determined				MD	Six miles south of Shingle Springs. Active prior to 1894. Quartz vein is black in appearance. (Crawford 94:104; 96:134.)
	Blue Bank	Not determined		11N	9E	MD	Nine miles northwest of Shingle Springs. Active in 1896. A 1½-ft. vein in amphibolite strikes north and dips 74° W. Developed by 120-ft. drift adit, 100-ft. inclined winze, and open cuts. Ore was treated in 2-stamp mill. (Crawford 96:134; Tucker 19:281.)
54	Blue Gouge (Berg)	Americo & Columbus Sciaroni, Grizzly Flat	21	10N	13E	MD	(Crawford 96:135; Logan 38:224-225; herein.)

LODE GOLD (CONT.)

MAP NO.	CLAIM, MINE, OR GROUP	OWNER NAME, ADDRESS	SEC.	T.	R.	B & M	REMARKS
	Blue Lead	Helen B. King, 20 Molly Lane, Placerville	3	11N	10E	MD	On Mother Lode one and one-half miles southeast of Garden Valley. Active around 1867. Specimen ore produced. Ore treated in 20-stamp mill. (Logan 34:21.)
	Board						See Bordt.
	Bollhalter	Not determined	1	10N	9E	MD	One mile south of Four Corners. Assays from large quartz outcrop in amphibolite near serpentine body were $1.40 to $3.80 per ton. (Logan 38:225.)
	Bona Forsa	Not determined		11N	10E	MD	Northwest of Placerville. Active around 1890. Two north-striking veins rich in sulfides. Developed by 50- and 65-ft. shafts and crosscuts. (DeGroot 90:177.)
55	Bonanza	Camilla D. Heald, 624 S. 13th Street, San Jose	11	8N	10E	MD	On Mother Lode 1 mile south of Nashville. Active 1922. Ore treated in 5-stamp mill. (Logan 23a:44.)
56	Boneset	Mrs. W. Moore, 1120 - 24th Street, Sacramento	10,11	10N	9E	MD	Three miles northwest of Rescue. Active in 1894; intermittently active in 1930's with a small production. A 10- to 15-ft. vein in metadiorite and greenstone strikes northeast and dips 70° northwest; free gold is in thin bands with pyrite, chalcopyrite, and galena. Developed by 140-ft. crosscut adit and several hundred feet of drifts and open cuts. (Crawford 94:104; 96:135; Tucker 19:282.)
	Bordt (Board)	Not determined	7	12N	10E	MD	Half a mile east of Greenwood. Active in 1894 and 1926.

LODE GOLD (CONT.)

MAP NO.	CLAIM, MINE, OR GROUP	OWNER NAME, ADDRESS	SEC.	T.	R.	B & M	REMARKS
	Bordt (continued)						A 4-ft. west-striking, north-dipping vein in slate. Developed by a 60-ft. drift adit. (Crawford 94:104; 96:135; Tucker 19:282.)
57	Boulder (Kaeser)	Not determined	32	11N	9E	MD	Five miles northwest of Rescue. Active in 1896; some activity in 1931 and 1934. A 7- to 8-ft. vein in granodiorite strikes northeast and dips 35° northwest. Developed by four adits, one a 400-ft. crosscut; extensive drifting. Ore was treated in a 10-stamp mill. (Crawford 96:135-136; Tucker 19:282; Logan 35:22; 38:234-235.)
	Brandon	Not determined	31	9N	10E	MD	Three and one-half miles northeast of Latrobe. Active many years ago. A north-trending body of greenstone containing disseminated auriferous pyrite; developed by shaft and adit.
58	Bret Harte (Safeguard No. 1)	Not determined	6	11N	12E	MD	On north slope of Slate Mountain four miles southwest of Pino Grande. Active around 1934 when several hundred tons of ore were mined from surface. There are two veins, one developed by 240-ft. adit and the other by a 50-ft. adit. (Logan 38:225.)
59	Briarcliffe (Baldwin, Last Chance)	Nick Neilsen, Box 100, Diamond Springs	1,2, 11,12	8N	10E	MD	(Crawford 94:102; 96:133; Tucker 19:280; Logan 38:226; herein.)

LODE GOLD (CONT.)

MAP NO.	CLAIM, MINE, OR GROUP	OWNER NAME, ADDRESS	SEC.	T.	R.	B & M	REMARKS
60	Bright Hope	Hope Oehm, 24-A Second Street, Woodland	2	12N	10E	MD	One mile northeast of Georgetown. Active in 1890 and 1896. A 6-ft. vein in Mariposa slate strikes northeast and dips 35° northwest. Developed by 400-ft. crosscut adit and 80-ft. shaft. (De Groot 90:177; Crawford 94:105; 96:136; Tucker 19:282.)
	Brooklyn		2,3	12N	10E	MD	See Beebe.
	Brown Bear		36	11N	10E	MD	See Placerville Gold Mining Company. (Irelan 88:182.)
61	Brust (Asbestos, Gold Hill)	William Hodge, c/o Margaret Hodge Rains, 932 Fresno Avenue, Berkeley 7	6	11N	11E	MD	One mile northeast of Spanish Flat. Active in 1926. Free gold is in calcite, quartz, and talc stringers in serpentine. Developed by several shallow shafts and 800-ft. adit. (Logan 23a:44; 26:412.)
62	Buena Vista	Blanch Schuster, P.O. Box 43, El Dorado	13,24	9N	10E	MD	On Mother Lode three miles southeast of El Dorado. Active around 1900; prospected 1936 and 1940-42 with very little production. Several veins are in schist. Developed by 208-ft. shaft with 400 ft. of drifts and 250-ft. adit. (Storms 00:91; Logan 38:227.)
	Bullard						See Golden Trace.
	Calaveras	Not determined		8N	10E	MD	Four miles east of Latrobe.· Active in 1896. A north-east-striking vein in slate and greenstone; contains free gold and auriferous sulfides. Developed by 32- and ·53-ft. shafts and open cuts. (Crawford 96:136.)

LODE GOLD (CONT.)

MAP NO.	CLAIM, MINE, OR GROUP	OWNER NAME, ADDRESS	LOCATION SEC.	T.	R.	B & M	REMARKS
63	Caledonia	Wayne Huckaby, Garden Valley	15	11N	10E	MD	Two miles west of Kelsey. Active around 1900; prospected in 1948-49 but no production. A 5-ft. vein contains auriferous pyrite. Developed by shaft.
64	California Consolidated (Ibid, Tapioca)	Albert N. Brown et al., 5252 James Avenue, Oakland	16	9N	13E	MD	One mile southwest of Grizzly Flat. Active in 1896; reopened in 1938. Several veins in mica schist including the Tapioca which strikes north 6° east, and dips northwest. Tapioca vein yielded ore containing $11.30 per ton. Developed by 468- and 70-ft. crosscut adits. Ore was treated at Morey mill. (Crawford 96:159; Logan 26:412-413; 38:227-228.)
65	California Jack	Not determined	15,22	12N	10E	MD	Three miles southwest of Georgetown. Active prior to 1896. A 12-ft. vein in Mariposa slate strikes north and dips 60° E. Developed by 350-ft. crosscut adit, 200-ft. north drift, and 90-ft. shaft. Ore was treated in 10-stamp mill. (Crawford 96:136; Tucker 19:282.)
	Central						See Inez.
	Chanced Upon						See Darling.
66	Chaparral (Golden Queen)	P.C. Stingle, c/o Lynn, Woodworth, and Evarts, 75 Federal Street, Boston, Mass.	26	11N	10E	MD	On Mother Lode two miles southwest of Kelsey. Active 1872-75, and 1901. A 6-ft. vein strikes northwest and dips 70° east with diabase footwall and slate hanging wall. Ore yielded $7 to $15 per ton. Developed by 200-ft. shaft and 50-ft. adit. Ore was treated in mill on property. (Tucker 19:283; Logan 34:21.)

LODE GOLD (CONT.)

MAP NO.	CLAIM, MINE, OR GROUP	OWNER NAME, ADDRESS	SEC.	T.	R.	B & M	REMARKS
	Chester		18	10N	11E	MD	See Placerville Gold Mining Company. (Irelan 88:182-183; DeGroot 90:173; Crawford 94:106.)
67	China Hill	E. J. Murray, Box 1282, Carmel	16	9N	10E	MD	Three miles southwest of El Dorado. Active prior to 1894. A 5-ft. vein in hornblende porphyry strikes north and dips 70° east. Ore is in small, rich shoots. Developed by 200-ft. crosscut adit, 200 ft. of drifts, and open cuts. Ore was treated in a 5-stamp mill. (Crawford 96:106; 96:137; Tucker 19:283.)
68	Church	Madre de Oro, Gold Mines Company, Box 925, Corcoran	12	9N	10E	MD	(Hanks 86:43; Irelan 88:191-193; DeGroot 90:171; Crawford 94:106; 96:137-138; Storms 00:92; Tucker 19:283; Logan 22:209; 26:413; 34:21-22; herein.)
	Cincinnati	Florence G. Kyle, Kelsey	3	11N	10E	MD	On Mother Lode, one and one-half miles southeast of Garden Valley. Active 1917-18. Gold occurs in stringer in decomposed dike in Mariposa slate. Ore yielded $3.82 per ton by amalgamation. Developed by open cuts and shallow shafts. Ore was treated in 5-stamp mill. (Logan 34:22.)
	Cinnamon Bear		36	11N	10E	MD	See Placerville Gold Mining Co. (Crawford 94:106; Irelan 88:182; Preston 93:201.)
	Clark						See Black Oak.

LODE GOLD (CONT.)

MAP NO.	CLAIM, MINE, OR GROUP	OWNER NAME, ADDRESS	LOCATION SEC.	LOCATION T.	LOCATION R.	LOCATION B & M	REMARKS
69	Coe Hill (Bathurst, Gold Star)	Not determined	33	12N	10E	MD	On Mother Lode one mile south of Garden Valley. Active 1921, 1925-1926. Several veins yielded $6 to $20 per ton. Developed by shallow shafts and an adit. Ore was treated in a 2-stamp mill. (Logan 26:417; 34:22-23.)
	Collins and Bacchi	C. S. Collins, Box 231, Placerville	28	12N	10E	MD	Near Garden Valley. Prospected prior to 1914. (Tucker 19:283.)
70	Cosumnes (Melton, Middle End)	Caldor Lumber Company, Diamond Springs	4	9N	13E	MD	(Irelan 88:178-180; Crawford 94:117; 96:150; Tucker 19:291; Logan 26:418; 38:238; herein.
71	Cousin Jack	Martha Agnes Weber (3/4), Placerville and Maryland Casualty Company (1/4), 240 Sansome Street, San Francisco	29	9N	13E	MD	Five miles southwest of Grizzly Flat. Active prior to 1894. A 1- to 4-ft. vein in slate strikes north and dips 60° west. Developed by 400- and 300-ft. drift adits and 70-ft. winze. (Crawford 94:107; 96:138; Tucker 19: 283.)
	Crescent						See Placerville Gold Mining Company.
72	Crown Point Consolidated (Bald Eagle, Gold Queen)	J. T. Richards Estate, Placerville	31	10N	11E	MD	On Mother Lode one and one-half miles southeast of Diamond Springs. Active 1894 and again around 1923. Three veins 4 to 20 ft. wide in Mariposa slate strike N. 27° E. and dip 72° east. Small lots of ore yielded $20 per ton or more. Developed by 500-ft. inclined shaft with 100-, 200-, 300-, and 400-ft. levels. A 600-ft. drain tunnel intersects shaft at 300-ft. level. A 150-ft. shaft south of main shaft. (Crawford 94:107; 96:138; Tucker 19:283-284; Logan 24:9; 34:23.)

LODE GOLD (CONT.)

MAP NO.	CLAIM, MINE, OR GROUP	OWNER NAME, ADDRESS	SEC.	T.	R.	B & M	REMARKS
	Crusader	Mary Ann Simpson, c/o J. P. Conroy, Box 212, El Dorado	12	9N	10E	MD	On Mother Lode two miles south of Diamond Springs. Active prior to 1914 and again in 1929. A 3-ft. vein in Mariposa slate strikes north 20° east and dips 80° west. Developed by 100-ft. inclined shaft with 100-ft. level. (Tucker 19:284.)
73	Crystal	W. L. and S. S. Lovejoy, Cool	18	12N	9E	MD	One-half mile north of Cool. Active in 1896; prospected in 1931. Vein in quartz prophyry strikes north and dips 15° west. Contains chalcopyrite. Developed by 25-ft. shaft and two 60-ft. inclines. (Crawford 96:138; Tucker 19:284.)
74	Crystal	Arthur S. Morey et al., 91 Ricon Way, San Francisco	32,33	9N	13E	MD	On Cosumnes River, five miles southeast of Grizzly Flat. Active around 1894. There are three nearly west-striking veins in granite with steep dips. Developed by 70- and 250-ft. shafts and 1200-ft. crosscut adit. Ore was treated in an 8-stamp mill. (Crawford 94:107; 96:138; Tucker 19:284.)
75	Crystal (El Dorado Crystal)	F. W. Barrette, Shingle Springs	18	9N	10E	MD	(DeGroot 90:178; Crawford 94:107-108; 96:138; Tucker 19:284; Logan 39:228-229; herein.)
76	Dailey and Bishop	Lucian Vaira, Drytown	27	9N	13E	MD	Two miles south of Grizzly Flat. Active around 1896. A 1½- to 3-ft. vein in slate strikes north. Developed by 800-ft. drift adit, crosscuts, and winze. Ore was treated in a 10-stamp mill. (Crawford 94:108; 96: 138-139; Tucker 19:284.)

LODE GOLD (CONT.)

MAP NO.	CLAIM, MINE, OR GROUP	OWNER NAME, ADDRESS	LOCATION					REMARKS
			SEC.	T.	R.	B & M		
77	Dalmatia (Kelly)	W.F.I. Bell Estate, c/o Jay R. Fogal, Box 100, Garden Valley	13	11N	10E		MD	Just east of Kelsey. Active 1880's, 1890-94 and again around 1935. Quartz seams and vein are in zone 20 to 50 ft. wide that strikes north 10° west and dips north-east. A 2-ft. vein contained $16 per ton; a pocket yielded $14,000; the seams yielded $2 to $3 per ton. Originally worked in open cut 500 ft. long. Developed by 200-ft. inclined shaft and 1200-ft. adit. Ore treated in Huntington mills and 10-stamp mill. (Irelan 88:177; DeGroot 90:174-175; Preston 92:201-202; Crawford 94:108; 96:139; Tucker 19:284; Logan 34:23.)
78	Darling (Chanced Upon)	E. V. Rhodes, Kelsey	33	12N	11E		MD	Four miles northeast of Spanish Flat. Active around 1892. A 2-ft. vein in slate strikes north and dips east and contains $5 to $6 per ton in free gold with calave-rite and petzite. Developed by 190-ft. shaft. Ore was treated in 10-stamp mill with Frue concentrators. (Preston 92:202; Crawford 94:108; 96:139; Tucker 19: 284.)
	Darrow							See Log Cabin.
	Davenport	Dudley G. Davenport et al., P.O. Box 33, Garden Valley	34	12N	10E		MD	On Mother Lode one-half mile east of Garden Valley. Active in 1934. Later, worked jointly with Black Oak Mine, which see. Gold occurs in amphibolite schist. Developed by 280-ft. crosscut adit and open cuts. (Logan 34:21.)
	Davey							See Black Oak.

LODE GOLD (CONT.)

MAP NO	CLAIM, MINE, OR GROUP	OWNER NAME, ADDRESS	SEC.	T.	R.	B & M	REMARKS
79	Davidson	Jerome M. Strickland et al., Route 2, Box 56, Placerville	22,27	10N	10E	MD	Two miles northwest of El Dorado. Originally active prior to 1894; prospected in 1947-48. A 2-ft. vein in slate strikes north 10° west and dips 74° northeast. Developed by 280-ft. inclined shaft with 100- and 300-ft.levels. Ore originally treated in 20-stamp mill and later in 5-stamp mill. (Crawford 94:108; 96:139; Tucker 19:284-285.)
	Dayton Consolidated						See Black Oak. (Logan 38:224,363.)
	Defiance	Not determined		10N	10E	MD	Five miles northeast Shingle Springs. Gold occurs in green schist. (Crawford 94:108.)
	Diamond						See Tullis.
	Diamond Springs						See Larkin.
	Donozo	Not determined		12N	10E	MD	One-half mile east of Greenwood. Vein strikes north 40° west and dips 55° northeast. Developed by 60-ft. drift adits. (Crawford 94:108; 96:139.)
	Duncan and Adams	Not determined	1,2	9N	10E	MD	One mile southeast of El Dorado. Active 1931 when 700 tons ore mined that yielded $10,266.
80	Eagle	Austin E. Borcham et al., Camp Seco	9	9N	13E	MD	One and one-half miles north of Grizzly Flat. Active in 1896. A 3-ft. vein in granodiorite strikes north-east and dips 75° northwest. Ore shoot 150 ft. long and as much as 6 ft. wide contains appreciable amounts

LODE GOLD (CONT.)

MAP NO.	CLAIM, MINE, OR GROUP	OWNER NAME, ADDRESS	LOCATION				REMARKS
			SEC.	T.	R.	B & M	
	Eagle (continued)						of pyrite, galena, and sphalerite. Developed by 780-ft. drift adit and 240-ft. shaft. (Irelan 88:178; Crawford 96:139; Tucker 19:285.)
	Eagle						See Ohio.
81	Eagle King	Marian S. Clark, c/o Eagle King Mining Company, 209 Edgewater, Balboa	4,9	9N	13E	MD	Two miles north of Grizzly Flat. Active 1894-96. A north to northeast-striking vein with 75° west to vertical dip is on or near granodiorite-mica schist contact. Vein is three to four feet wide and contains appreciable amounts of pyrite, galena, and sphalerite. Developed by 1200-ft. drift adit and 60-ft. winze 200 ft. from portal. Ore was treated in 10-stamp mill. (Irelan 88:178; Crawford 94:108-109; 96:139; Tucker 19:285.)
	East Lode						See Beebe.
	Edmunds						See No. 2.
	Edner	D. K. Moore, Omo Ranch	8	8N	13E	MD	One and one-half miles southeast of Omo Ranch. Active in 1896. A 1½-ft. vein in granodiorite strikes north and dips 65° west. Developed by 150-ft. adit and 50-ft. shaft. (Crawford 94:109; 96:139-140; Tucker 19:285.)

LODE GOLD (CONT.)

MAP NO.	CLAIM, MINE, OR GROUP	OWNER NAME, ADDRESS	SEC.	T.	R.	B & M	REMARKS
	El Dorado Big Tunnel Company			11N	11E	MD	During the 1890's this concern operated a mine at Big Canyon two miles north of Placerville which later was purchased by the Gentle Annie Mining Company. (See also Gentle Annie and River Hill Group.)
	El Dorado Crystal						See Crystal.
	Elliott (Sir Water Raleigh)	Not determined	19	10N	11E	MD	On Mother Lode two miles south of Placerville. Active around 1894. A 4-ft. vein in slate strikes northwest and dips 80° northeast. Developed by 50-ft. inclined shaft and crosscut adit. (Crawford 94:109; 94:140.)
	Emma	Mabel H. Poheim, 598 Bush Street, San Francisco	21,28	12N	10E	MD	Two miles northwest of Garden Valley on Mother Lode. Active prior to 1890. Developed by 100-ft. shaft. (DeGroot 90:176; Crawford 96:140.)
	Empire Group						See Polar Bear.
82	Epley Consolidated (Mammoth)		20	10N	11E	MD	Also see Placerville Gold Mining Company. (Irelan 88: 180-187; DeGroot 90:173; Crawford 96:109.)
	Equator	Not determined		10N	11E	MD	Three miles south of Diamond Springs on Mother Lode. Active 1888-90. Three veins in slate strike northeast and dip southeast. Developed by 1300-ft. crosscut adit and 110-ft. inclined shaft. (Irelan 88:190; DeGroot 90:172; Crawford 94:109.)
	Esperanza						See Skipper.

LODE GOLD (CONT.)

MAP NO.	CLAIM, MINE, OR GROUP	OWNER NAME, ADDRESS	SEC.	T.	R.	B & M	REMARKS
	Esperanza (Garden Valley)	Not determined	28	12N	10E	MD	One mile northwest of Garden Valley. Active 1890-1900. A northwest-trending vein associated with lens of amphibolite in slate developed by 600-ft. vertical shaft and 700 feet of drifts. Ore treated in 20-stamp mill. (DeGroot 90:175; Crawford 94:109; 96:140; Tucker 19: 285; Logan 34:23-24.)
	Eureka		11	12N	10E	MD	See Woodside-Eureka and Beebe.
	Eureka	L. A. Frontz, Georgetown	2,11	12N	10E	MD	In Georgetown. Active prior to 1888. Three parallel veins in Mariposa slate strike northeast and dip 60° southeast. Veins are 6-10 ft. wide. Developed by 240-ft. inclined shaft and 500 feet of drifts. See also Beebe. (Irelan 88:182; DeGroot 90:178; Preston 92: 203; Crawford 94:109; Tucker 19:285.)
	Eureka		6 / 36	10N / 11N	11E / 11E	MD / MD	See Placerville Gold Mining Company. (DeGroot 90:178; Preston 92:203; Crawford 94:109; Tucker 19:285.)
	Excelsior Consolidated						See Garfield and Excelsior Consolidated.
83	Expansion	Life Estate of Josie G. Pine, c/o John Charles Pine et al., Route 2, Box 34, Placerville	17	10N	10E	MD	Three miles north of Shingle Springs. Active 1900-04; prospected in 1936. Disseminated auriferous pyrite in amphibolite schist. Developed by 150-ft. crosscut adit. (Logan 38:229.)

LODE GOLD (CONT.)

MAP NO.	CLAIM, MINE, OR GROUP	OWNER NAME, ADDRESS	LOCATION				REMARKS
			SEC.	T.	R.	B & M	
84	Falls	Mary A. Simpson et al., Box 212, El Dorado	1	9N	10E	MD	On Mother Lode two miles south of Diamond Springs. Active in 1914 and again around 1934. Vein in Mariposa slate. Developed by 235-ft. crosscut adit. (Tucker 19:285; Logan 34:24.)
	Faraday	Not determined	20,17, 18	10N	11E	MD	See Placerville Gold Mining Company. (Irelan 88:186; DeGroot 90:173; Crawford 94:110.)
	Flagstaff	Not determined					A few miles north of Grizzly Flats. Active around 1888. Ore treated in 10-stamp mill. (Irelan 88:110; Crawford 96:178.)
	Fort Yuma	Capital Company, 1 Powell Street, San Francisco 2	29, 32	9N	10E	MD	On Big Canyon Creek two miles northeast of Brandon Corner. Active 1890-1902 and again in 1938. A 2- to 4-ft. vein in Calavera's slate strikes north and dips east. Developed by 175 and 40-ft. shaft and drifts. Ore treated by flotation . (Crawford 94:110; 96:140; Logan 38:229.)
85	French Creek	William Lange and L.W. Loomis, Placerville	31	9N	10E	MD	Herein.
86	Frog Pond and Marigold Consolidated (Marigold Consolidated)	Not determined	28	12N	10E	MD	One-half mile northwest of Garden Valley. Active intermittently 1914-27. Flat gold-bearing seams with arsenopyrite in amphibolite. Developed by 60-ft. shaft, drift. Ore was treated in 2-stamp mill. (Tucker 19: 286; Logan 38:230.)

LODE GOLD (CONT.)

MAP NO.	CLAIM, MINE, OR GROUP	OWNER NAME, ADDRESS	LOCATION SEC.	LOCATION T.	LOCATION R.	LOCATION B & M	REMARKS
87	Funny Bug (Pendelco)	R. V. Montgomery, Route 2, Box 193-K, Placerville	5	10N	10E	MD	(Logan 38:242-243; Eric 48:230; herein.)
	Gamblin						See Gambling.
88	Gambling (Gamblin)	Jean E. Carlson, 211 Palomares Drive, Ventura	5	8N	12E	MD	Two miles southwest of Fairplay. Active 1915-18 and 1933-34. An 18- to 30-inch vein in granodiorite strikes west and dips 80° south. Developed by 500-ft. inclined shaft, considerable drifting, and adit on 90-ft. level. Ore was treated in 45-ton mill with 10 stamps, plates, and Frue vanners. (Logan 35:23; 38:230-231.)
	Garden Valley						See Esperanza.
	Gardner Consolidated	W. H. Myers, 24 Bedford Avenue, Placerville	6	10N	11E	MD	One mile north of Placerville. Active prior to 1914. A 5-ft. vein in Mariposa slate strikes north 40° east and dips 70° east. Developed by 400-ft. crosscut adit and 500 ft. of drifts. (Tucker 19:286.)
	Garfield	Not determined	18 or 19	13N	11E	MD	One mile south of Volcanoville. Active around 1894. vein in slate strikes north and dips 55° west. Developed by 120-ft. inclined shaft and 700-ft. crosscut adit. (Crawford 94:110; Tucker 19:286.)
89	Garfield and Excelsior Consolidated	Not determined	5,6	12N	10E	MD	One mile northeast of Greenwood. Active around 1894. A 20-ft. vein in slate with mineralized footwall. Developed by four crosscut adits, 200 to 400 ft. long. (Crawford 94:111; Tucker 19:286.)

LODE GOLD (CONT.)

MAP NO.	CLAIM, MINE, OR GROUP	OWNER NAME, ADDRESS	LOCATION				REMARKS
			SEC.	T.	R.	B & M	
	Garibaldi Consolidated	Not determined					Near Greenwood. Prospected around 1894. A 6-ft. vein in slate strikes northwest and dips northeast. (Crawford 94:111.)
	Gentle Annie		6	10N	11E	MD	See River Hill. (De Groot 90:177; Crawford 94:111; 96:141-143; Storms 00:94; Logan 34:45.)
	Georgia Slide						See Seam gold section.
90	German (Haeger)	Hazel A. Hill, Box 12, El Dorado	14	9N	10E	MD	On Mother Lode three miles south of El Dorado. Active around 1896-1900. Vein in Mariposa slate strikes north and dips 60° east. Developed by 500-ft. inclined shaft. Ore was treated in 10-stamp mill. (Crawford 96:142; Storms 00:90; Tucker 19:286.)
	Gloriana	Not determined	29	9N	13E	MD	At Henry's Diggings two miles north of Omo Ranch. Long idle.
	Gold Coin						See Veerkamp.
	Golden Gate						See McNulty.
	Golden Oak						See Big Four.
	Golden Queen						See Chaparral.

LODE GOLD (CONT.)

MAP NO.	CLAIM, MINE, OR GROUP	OWNER NAME, ADDRESS	LOCATION				REMARKS
			SEC.	T.	R.	B & M	
	Golden Trace (Bullard)	Not determined		9N	13E	MD	Two and one-half miles north of Grizzly Flat by North Fork Cosumnes River. Active prior to 1896. A 2½-ft. vein in granite trends north and dips west. Developed by 250-ft. drift adit. (Crawford 96:143.)
	Golden Unit						See Argonaut.
	Gold Hill						See Brunt.
	Gold Note	Not determined	4	8N	13E	MD	Two miles southeast of Omo Ranch. Long Idle.
	Gold Note						See Philadelphia and Gold Note.
	Gold Queen						See Crown Point Consolidated.
	Gold Reserve						See Pyramid.
	Gold Star						See Coe Hill.
	Good Luck	Not determined		10N	11E	MD	Two miles east of Diamond Springs. Active 1909-10. An 18-inch vein in Mariposa slate strikes north 20° east and dips 45° east. Developed by 250-ft. shaft and 200- and 300-ft. adits. Ore treated in 5-stamp mill. (Tucker 19:287.)
	Good Luck						See Jones.

LODE GOLD (CONT.)

MAP NO.	CLAIM, MINE, OR GROUP	OWNER NAME, ADDRESS	LOCATION SEC.	T.	R.	B & M	REMARKS
91	Gopher-Boulder	W. A. Bell, Box 100, Garden Valley	11,14	11N	10E	MD	On Mother Lode one mile northwest of Kelsey. Active 1858 when $15,600 produced; $40,000 produced in 1880's; prospected in 1931 and again during 1934-36. Two veins in slate and greenstone up to 50 ft. in width; Gopher vein strikes north 23° west; Boulder vein strikes north 11° west; both dip northeast. Ore averaged $2.50 to $6 per ton; a few areas contained as much as $16 per ton. Developed by 260-ft. inclined shaft with levels at 50-ft. intervals, a 200-ft. drift adit, 850-ft. crosscut adit, and open cuts. Ore was treated in 20-stamp mill and two Huntington mills. (Irelan 88:175-177; Crawford 96:143; Storms 00:98; Tucker 19:287; Logan 34:24-25.)
92	Grand Victory	Provident Minerals Corporation, c/o Floyd Singleton, 459 Turk Street, San Francisco 2	33,34	10N	11E	MD	(Irelan 88:194; DeGroot 90:178; Crawford 94:112; 96: 143; Tucker 19:287; Logan 35:23; 38:231-232; herein.)
	Gray (Old Gray)	Not determined		10N	9E	MD	Three miles northwest of Shingle Springs. Active about 1894. A 1- to 3-ft. vein strikes west, dips 45° north, and is near gabbro body. Developed by 100-ft. shaft and drift. (Crawford 94:119; 96:143.)
	Green	Not determined	19	13N	11E	MD	A chromite mine originally mined for gold. See in Chromite section.
	Greenstone						See Barnes-Eureka

LODE GOLD (CONT.)

MAP NO.	CLAIM, MINE, OR GROUP	OWNER NAME, ADDRESS	LOCATION				REMARKS
			SEC.	T.	R.	B & M	
93	Greenwood	Not determined	12,13	12N	9E	MD	At Greenwood. Intermittently active 1937-40. Gold occurs in seams across width of 200-ft. open pit. Ore treated in mill with ball mills and cyanidation. (Logan 38:232.
94	Griffith Consolidated	Frank B. Richards et al., Box 611, Placerville	30,31	10N	11E	MD	On Mother Lode one-half mile southeast of Diamond Springs. Worked originally in 1850's, active 1888-90, 1896, and 1903. A 5- to 12-ft. vein in slate strikes northeast and dips southeast. Ore yielded from $4.25 to $8 per ton, although in a few areas it yielded as much as $65 per ton. Developed by 700-, 150-, and 253-ft. shafts and drifts. Ore was treated in 5-stamp mill. (Irelan 88:189; DeGroot 90:172; Crawford 94:112; 96:144; Storms 00:92-93; Tucker 19:287; Logan 34:25-26.)
95	Grit (Liddicoat, Spanish Dry Diggings)	Liddicoat Gold Mines Company, Inc., c/o Joseph L. Liddicoat, Route A, Box 27, Greenwood	30	13N	10E	MD	(Logan 20:426; 23a:29; 23b:44; 24:8; 26:414; 34: 44,46; herein.)
96	Gross Consolidated		6	10N	11E	MD	See Placerville Gold Mining Company. (Irelan 88:181-182; DeGroot 90:173; Crawford 94:113; 96:161.)
	Grouse Gulch	Not determined	16	9N	13E	MD	One and one-half miles west of Grizzly Flat. Active prior to 1888. A vein in granitic rock strikes north and dips 60° west, and ranges from ½ ft. to 5 ft. in width. Development by 100-, 80-, and 50-ft. shafts, drifts and 200-ft. drain tunnel. (Irelan 88:178;

LODE GOLD (CONT.)

MAP NO.	CLAIM, MINE, OR GROUP	OWNER NAME, ADDRESS	LOCATION				REMARKS
			SEC.	T.	R.	B & M	
	Grouse Gulch (continued)						Crawford 94:113; 96:144; Tucker 19:288.)
97	Guildford (Poverty Point)	A. A. McKinnon, Placerville	25,36 6	11N 10N	10E 11E	MD MD	On Mother Lode two miles north of Placerville. Active 1912-17, when more than $200,000 produced; intermittently active 1920-25; minor production 1931-32. Two parallel veins in Mariposa slate strike north 20° west and dip 70° southeast. Two ore shoots, 200 and 400 ft. long averaging 5 ft. in width, were mined. Ore yielded $4 to $5 per ton while pyrite concentrate ranged from $40 to $88 per ton. Developed by four drift adits, 500, 600, 700, and 1500 ft. long. Ore treated in 15-stamp mill with Wilfley tables and Frue vanners. (Tucker 19:287-288; Logan 26:414; Knopf 29:49; Logan 34:36-37.)
	Haeger						See German.
	Hall Consolidated						See River Hill group.
	Halleck						See Placerville Gold Mining Company.
	Hardscratch	Not determined		9N	13E	MD	Two miles west of Grizzly Flat. Active around 1896. A northeast trending vein in granite dips northwest. Developed by 120-ft. crosscut adit, 100-ft. drift, and 26-ft. shaft. Ore treated in small Dyer mill. (Crawford 96:144.)

LODE GOLD (CONT.)

MAP NO.	CLAIM, MINE, OR GROUP	OWNER NAME, ADDRESS	SEC.	T.	R.	B & M	REMARKS
98	Harman group		6,7, 36	10N 11N	11E 10E	MD MD	See Placerville Gold Mining Company. (Irelan 88:181-182; DeGroot 90:173,178; Preston 92:203; Crawford 94: 109, 113; 96:161; Tucker 19:285; Logan 34:26.)
	Havilah						See Nashville.
99	Hazel Creek	Fay M. R. Gunby, Box 752, Placerville	3	10N	13E	MD	(Herein.)
		Leased by Hazel Creek Mining Company, P.O. Box 508, North Sacramento					
100	Henrietta		20	10N	11E	MD	See Placerville Gold Mining Company. (Irelan 88:186; DeGroot 90:173; Crawford 94:113.)
	Hidden Treasure	Not determined	8	8N	13E	MD	One mile northeast of Indian Diggins. Long idle.
	Hillside Group						See Martinez.
101	Hines-Gilbert	Wayne Huckaby, 201 South 8th Street, Sacramento	23,24	13N	9E	MD	On north end Mother Lode one mile northwest of Spanish Dry Diggings by American River. Active 1921-28, 1934; prospected in 1954. Two northwest-striking quartz veinlets occur in slate and amphibolite schist. Developed by open cut 150 ft. in width and 450-ft. adit.

LODE GOLD (CONT.)

MAP NO.	CLAIM, MINE, OR GROUP	OWNER NAME, ADDRESS	LOCATION				REMARKS
			SEC.	T.	R.	B & M	
	Hines-Gilbert (continued)						Ore treated in 10-stamp mill. (Logan 21:427; 23a:209; 24:178-179; 26:414; 34:27-28, 46.)
	Holly	Not determined	7	10N	11E	MD	In Placerville. Active in 1918.
	Homestead	Not determined	26	13N	10E	MD	Two and one-half miles north of Georgetown. Active 1912-13.
	Hope		16	9N	13E	MD	South extension of Mount Pleasant mine. (Crawford 94:113.)
	Humbug						See Morey.
	Humphrey	Not determined		8N	12E	MD	Two and one-half miles southwest of Fairplay. Active around 1896. A 16- to 24-inch vein in granite strikes west and dips 80° south. Contains pyrite and galena. Developed by 15-ft. shaft and 70-ft. drift. (Crawford 96:145.)
	Hustler						See Adjuster.
	Ibid		21	9N	13E	MD	See also California Consolidated. (Crawford 94:113; 96:145.)
	Ida		18	10N	11E	MD	See Placerville Gold Mining Company. (Irelan 88:182-183.)

LODE GOLD (CONT.)

| MAP NO. | CLAIM, MINE, OR GROUP | OWNER NAME, ADDRESS | LOCATION | | | | REMARKS |
			SEC.	T.	R.	B & M	
	Idaho	Hazel A. Hill, Box 12, El Dorado	14,23	9N	10E	MD	Three miles south of El Dorado; long idle. (Tucker 19:288.)
102	Ida Livingston	Ida Livingston Gold Mining Company, 413 High Street Klamath Falls, Oregon	13	11N	10E	MD	On Mother Lode one mile north of Kelsey. Active prior to 1914. A 25-ft. vein in Mariposa slate strikes north 50° east and dips 80° southeast. Ore yielded up to $26 per ton. Developed by 150-ft. shaft. (Tucker 19:288; Logan 34:28.)
	Idlewild						See Taylor.
	Independence	Starlight Mining Co., 2626 Baker Street San Francisco	10	9N	10E	MD	A pocket mine two miles southwest of El Dorado. Active prior to 1914. (Tucker 19:288.)
	Independence	Not determined	27	12N	11E	MD	Four miles northwest of Slate Mountain. Active in 1933; ore treated in 2-stamp mill.
103	Inez (Central	M. B. Maginess, 1400 Vilvo Street San Francisco	1	8N	10E	MD	One mile east of Nashville. Active around 1890. Vein in slate strikes northwest and dips 60° northeast. Developed by 250-ft. shaft. (DeGroot 90:171; Crawford 94:114; 96:147; Tucker 19:288.)
	Iowa						See Beebe.

LODE GOLD (CONT.)

| MAP NO. | CLAIM, MINE, OR GROUP | OWNER NAME, ADDRESS | LOCATION | | | | REMARKS |
			SEC.	T.	R.	B & M	
104	Isabel (Isbell)	Helen B. King, 20 Molly Lane, Placerville	3	11N	10E	MD	On Mother Lode one mile southeast of Garden Valley. Active prior to 1914. A 2- to 8-ft. vein in slate strikes northwest. Developed by open cuts and 30-ft. shaft. Ore was treated at the Blue Lead Mill. (Tucker 19:288-289; Logan 34:28.)
	Isbell						See Isabel.
	Ivanhoe	Not determined	28	12N	10E	MD	One-half mile northwest of Garden Valley. Active prior to 1890. A vein in slate and greenstone strikes northwest and dips northeast. Developed by open cuts and 200-ft. shaft. (DeGroot 90:175-176; Crawford 94:114; Tucker 19:289.)
	James Marshall						See Big Sandy.
	Jones (Good Luck)	Not determined	6	9N	11E	MD	Two miles south of Diamond Springs. Active in 1915 and during 1922-23 when several thousand dollars produced. Vein strikes west. Developed by shaft with levels at 75, 165, and 225 feet and drifts. (Logan 23a:45; 23b:142.)
105	Josephine	J. A. Shields and Lester Marks, Auburn	5,6,7,8	13N	11E	MD	At Volcanoville. Active 1889-90, 1896, 1920, and 1934-35. A 6-ft. vein with ribbon structures in slate and serpentine strikes northeast and dips 45° southeast. Ore shoots are up to several hundred feet in length. Developed by five drift adits. Ore was treated in 20-stamp mill. (Irelan 88:165-166; DeGroot 90:178;

LODE GOLD (CONT.)

MAP NO.	CLAIM, MINE, OR GROUP	OWNER NAME, ADDRESS	SEC.	T.	R.	B & M	REMARKS
	Josephine (continued)						Preston 92:206; Crawford 94:114; 96:147; Tucker 19:289; Logan 20:427.)
	Kaeser						See Boulder.
	Kelly	Not determined	13	11N	10E	MD	On Mother Lode one-half mile north of Kelsey. Active around 1902 and 1932. A 6-ft. vein in Mariposa slate strikes north. Developed by two 50-ft. shafts. (Logan 34:29.)
	Kelly						See Dalmatia.
106	Kelsey (Lady)	A.P.O. Crattree, 616 North F Street, Porterville	24,25	11E	10E	MD	(Tucker 19:289; Logan 34:29, 211; 38:235, 236; herein)
	Lady						See Kelsey.
	Lady Blanche	Not determined	36	9N	12E	MD	Three and one-half miles east of Fairplay. Active in 1896. A 1- to 4-ft. vein strikes north 20° west and dips 50° northeast with slate footwall and porphyry hanging wall. Developed by 180-ft. and 80-ft. adits. (Crawford 96:147; Tucker 19:289.)
107	Lady Emma	J. L. Peters, Kelsey	13	11N	10E	MD	One mile east of Kelsey. Active around 1896; prospected in 1942 and 1947. A 4-ft. vein in slate and porphyry strikes north and dips 60° east. Developed by 300-ft. inclined and 150-ft. vertical shafts with drifts and

LODE GOLD (CONT.)

MAP NO.	CLAIM, MINE, OR GROUP	OWNER NAME, ADDRESS	SEC.	T.	R.	B & M	REMARKS
	Lady Emma (continued)						crosscuts. Ore was treated in 10-stamp mill. (Crawford 96:147-148; Tucker 19:289.)
108	La Moille	Walter Bidstrup, Box 35, El Dorado	11	9N	10E	MD	Three miles south of El Dorado. Active prior to 1894. A number of quartz seams trend north. (Crawford 94:115; 96:148.)
109	Larkin (Diamond Springs)	Bernard B. Ball, 122 Bedford Avenue Placerville	29	10N	11E	MD	(Crawford 96:148; Storms 00:93; Lang 07:963; Aubury 08:217; Tucker 19:277,289; Logan 26:407-408; 34:30; 47:233; Eric 48:231; herein.)
	Last Chance	Not determined	19	13N	11E	MD	One mile south of Volcanoville. Active in 1896. A 5-ft. vein in slate and porphyry strikes west and dips 55° south. Developed by 50-ft. vertical shaft and 400-ft. crosscut adit. Ore treated in 4-stamp mill. (Crawford 96:148.)
	Last Chance						See Briarcliffe.
	Liddicoat						See Grit.
	Live Oak	Frank Richards et al., Box 611, Placerville	30	10N	11E	MD	One mile east of Diamond Springs. Active prior to 1896; a vein in slate and greenstone strikes north and dips east. Developed by 30-ft. shaft. (Crawford 96:149.)

LODE GOLD (CONT.)

MAP NO.	CLAIM, MINE, OR GROUP	OWNER NAME, ADDRESS	LOCATION SEC.	T.	R.	B & M	REMARKS
110	Log Cabin (Darrow)	T. E. Massie, c/o John Kessler, 550 Swanston Drive, Sacramento	30	9N	10E	MD	Five miles south of Shingle Springs. Active 1894-96. A 2- to 16-ft. vein in slate and serpentine strikes northwest and dips 65° northeast. Developed by 600-ft. crosscut adit, drifts, and 30-ft. shaft. Ore was treated in 5-stamp mill. (Crawford 94:116; 96:149; Tucker 19:289.)
111	Lone Jack	John W. Reiss et al., Garden Valley	28	12N	10E	MD	On Mother Lode one mile northwest of Garden Valley. Active 1890-96. A vein up to 24 ft. in width contained $6 per ton. Developed by 400-ft. shaft. Ore was treated in 10-stamp mill. (DeGroot 90:176; Crawford 94:116; 96:149; Tucker 19:290.)
112	Lone Star	Camilla D. Heald, 624 South 13th Street San Jose	11	8N	10E	MD	One mile southeast of Nashville between the forks of the Cosumnes River. Active around 1894. A 7-ft. vein in slate strikes north and dips 60° east. Developed by 100-ft. shaft and 100-ft. drift. Ore was treated in two 5-ft. Huntington mills. (DeGroot 90:178; Crawford 94:116; 96:149; Tucker 19:290.)
113	Lone Star	Frank B. Richards et al., Box 611, Placerville	31	10N	11E	MD	Two miles southeast of Diamond Springs. Active 1894-96 and 1907-08. A 2- to 5-ft. vein in slate and greenstone strikes northeast with vertical dip. Developed by 500-ft. crosscut adit. (Crawford 94:116; 96:149; Tucker 19:290.)
114	Lookout	Hazel A. Hill, Box 12, El Dorado	11	9N	10E	MD	On Mother Lode three miles southwest of El Dorado. Intermittently active from 1860 to 1930's; $2,200 produced in 1912, $15,000 in 1933. Gold is in small but rich ore shoots in vein in Mariposa slate. Developed

LODE GOLD (CONT.)

MAP NO.	CLAIM, MINE, OR GROUP	OWNER NAME, ADDRESS	SEC.	T.	R.	B & M	REMARKS
	Lookout (continued)						by 400-ft. adit. (Tucker 19:290; Logan 22:45; 34:30.)
	Lookout and K. K.	Not determined	19	13N	11E	MD	In Quartz Canyon near Volcanoville. Active 1894-96. A 2-ft. vein in slate and porphyry strikes north and dips east. Developed by 200-ft. crosscut adit, drifts, and 34-ft. inclined shaft. (Crawford 94:116; 96:149; Tucker 19:290.)
115	Loveless	Hazel Hill, Box 12, El Dorado	11,14	9N	10E	MD	Three miles south of El Dorado. Active around 1914. A 1-ft. vein in slate and diabase strikes north and dips 47° west. Ore occurs in pockets. Developed by 160-ft. crosscut adit, 300-ft. drift, and 90-ft. shaft. (Tucker 19:290.)
	Lucinda	Not determined		9N	13E	MD	Three miles west of Grizzly Flat. Active prior to 1896. A 6-in. to 3-ft. vein in granodiorite and slate strikes west and dips 70° north. Developed by 50-ft. vertical shaft and 150-ft. crosscut adit. (Crawford 96:149; Tucker 19:290.)
	Lucky Jack	Walter Bidstrup, Box 35, El Dorado	11	9N	10E	MD	Two miles south of El Dorado. Active around 1914. Series of north-striking veins in granodiorite. Developed by several shallow shafts. Ore was treated in 2-stamp mill. (Tucker 19:290.)
	Lucky Marion (Shepard)	Not determined	12	12N	9E	MD	On Mother Lode one-half mile west of Greenwood. Active 1896-97 and again in 1901 when $3,860 produced. An 18- to 24-inch vein strikes north 30° west and dips 56° northeast with serpentine hanging wall and slate foot-wall. Developed by 112-ft. inclined shaft. Ore treated in 20-stamp mill. (Crawford 96:150; Tucker 19:290; Logan 34:30.)
	Lucky Star						See River Hill.

LODE GOLD (CONT.)

MAP NO.	CLAIM, MINE, OR GROUP	OWNER NAME, ADDRESS	LOCATION SEC.	LOCATION T.	LOCATION R.	LOCATION B & M	REMARKS
116	Lukens	P.E. Bottorf, Route 4, Box 6, Placerville	25	12N	8E	MD	Three miles southwest of Cool. Active around 1923. A narrow vein in amphibolite yielded some high-grade ore. Developed by 60- and 90-ft. shafts connected by 150-ft. drift. (Logan 23b:142-143; 26:414-415.)
	Lyon						See River Hill.
	Madrona	Not determined	29	12N	10E	MD	Two miles northwest of Garden Valley. Active prior to 1894. Vein in slate contains high percentage of sulfides. Developed by 40-ft. shaft. (Crawford 94:117.)
	Maltby	Not determined	17	12N	10E	MD	On Mother Lode one and one-half miles southeast of Greenwood and just north of Argonaut mine. Prospected during 1930's. Developed by adit.
117	Mammoth	Dorothy F. Kipp, Coloma	3, 4	10N	9E	MD	One mile northwest of Deer Valley School. Opened in 1860 when $10,000 produced from pocket; dump worked in 1934. A 3- to 4-ft. vein in granodiorite strikes northeast and dips northwest while 5-ft. vein strikes west. Developed by 75-ft. crosscut adit and 120-ft. drift; ore treated in 10-stamp mill. (Crawford 96:150; Tucker 19:290; Logan 38:237.)
	Mammoth						See Epley Consolidated and Placerville Gold Mining Company.
	Manhatten-California						See Manhatten Consolidated.

LODE GOLD (CONT.)

MAP NO.	CLAIM, MINE, OR GROUP	OWNER NAME, ADDRESS	SEC.	T.	R.	B & M	REMARKS
				LOCATION			
118	Manhatten Consolidated (Manhatten-California)	T. N. Reiman, 1616 S. Tuxedo Avenue, Stockton	36	9N	10E	MD	Two miles northeast of Nashville. Active prior to 1915; prospected in 1935, 1947, and 1952. Developed by 400-ft. shaft. (Tucker 19:291.)
	Manzanita	Not determined	24	11N	10E	MD	On Mother Lode one mile south of Kelsey. Active in 1918.
	Manzanita Queen						Extension of Griffith Consolidated. (Irelan 88:189.)
119	Marguerite	Loren, Ralph, and Berthal Crain, and Wilma Berriman, Plymouth	20,29	10N	11E	MD	One mile east of Diamond Springs. Active prior to 1914. Three parallel veins in Mariposa slate strike north and dip 70° east. Developed by 300-ft. vertical shaft, 200-ft. adit, and 1200 ft. of drifts. (Tucker 19:291.)
	Marigold Consolidated						See Frog Pond and Marigold Consolidated.
120	Martinez (Hillside group)	Martinez Gold Mines Company, Post Office Box 61, El Dorado	12,13	9N	10E	MD	(Tucker 19:288; Logan 26:415; 34:30-31; 38:237-238; herein.)
	Maryland						See Placerville Gold Mining Company.
	Mathenas Creek (Schneider)	Sophia (Schainman) Haut, 740 - 35th Avenue, San Francisco 21	31	10N	11E	MD	One mile south of Diamond Springs. Active 1888-94. A 2- to 8-ft. vein in slate strikes north 20° west and dips 55° northeast. Developed by 100- and 300-ft. adits. Ore treated in Huntington mill.

LODE GOLD (CONT.)

MAP NO.	CLAIM, MINE, OR GROUP	OWNER NAME, ADDRESS	LOCATION SEC.	LOCATION T.	LOCATION R.	B & M	REMARKS
	Mathenas Creek (continued)						(Irelan 88:190; DeGroot 90:172; Crawford 94:117; 96:150; Tucker 19:291.)
121	McNulty (Golden Gate, Oakland)	Hazel A. Hill, Box 12, El Dorado	14,23	9N	10E	MD	Three miles south of El Dorado on Mother Lode. Active 1890-94. A 6-ft. vein strikes north and dips 60° east with slate hanging wall and greenstone footwall. Developed by 400-ft. shaft, 450-ft. crosscut adit, and 450-ft. winze sunk from adit. Ore was treated in 10-stamp mill. (DeGroot 90:171; Crawford 94:117; 96: 150; Tucker 19:296.)
	Melton						See Cosumnes. (Irelar 88:178-180; Crawford 94:117; 96:150; Tucker 19:291; Logan 38:238.)
	Middle End						See Cosumnes. (Logan 26:418.)
	Miller (Ribbon Rock)	Not determined		10N	11E	MD	On Mother L de two miles south of Placerville. Active 1888-94, 1900. A 2- to 5-ft. vein in slate strikes north 120 west and dips 65° northeast. Much ribbon structure. Developed by 400-ft. inclined shaft. (Irelan 88:189; DeGroot 90:172; Crawford 94:122; 96: 150; Storms 00:94; Tucker 19:291; Logan 34:31.)
	Mitchell	Not determined	7	10N	9E	MD	Two miles northwest of Pine Hill. A 4- to 10-ft. vein developed by 150-ft. adit. (Logan 38:238.)
122	Monarch-Sugar Loaf	Leased by George Ross, Placerville	27	9N	10E	MD	Herein.

LODE GOLD (CONT.)

MAP NO.	CLAIM, MINE, OR GROUP	OWNER NAME, ADDRESS	SEC.	T.	R.	B & M	REMARKS
	Montana	Not determined	19	13N	11E	MD	One mile south of Volcanoville. Active 1896. A 14-ft. vein in slate and porphyry strikes north and dips 50° east. Developed by 45-ft. inclined shaft and crosscut adit. Ore treated in 1-stamp mill. (Crawford 96:151; Tucker 19:291.)
	Montauk Consolidated						See Zantgraf.
123	Montezuma-Apex and Montezuma Extension	Montezuma-Apex Mining Company., c/o of National Tunnel and Mines Company, 818 Kearns Building Salt Lake City, Utah	35 2	9N 8N	10E 10E	MD MD	(Crawford 94:118; 96:151; Storms 00:91; Tucker 18:291; Logan 26:415-416; Huttl 35a:173-175; Logan 38:238-240; herein.)
	Montezuma Extension						See Montezuma-Apex.
124	Morey (Humbug)	E. W. M rey, Grizzly Flat	16	9N	13E	MD	One mile west of Grizzly Flat. Active intermittently from 1919-44. Small amounts of high-grade ore occur in small veins. Developed by numerous shallow shafts. Ore was treated in small mill on property. (Logan 38:241; Heyl and others 48:230.)
125	Morman Hill	Joseph J. Kaline, 2131 Lake Street, San Francisco 21	5	10N	9E	MD	Four miles northwest of Rescue. Active 1934, 1938-41. A 24- to 30-inch vein strikes north 40° east and dips northwest. Developed by 110-ft. inclined shaft with 80 ft. of drifts on 100-ft. level. Ore treated in 5-stamp mill. (Logan 38:241.)

LODE GOLD (CONT.)

MAP NO.	CLAIM, MINE, OR GROUP	OWNER NAME, ADDRESS	SEC.	T.	R.	B & M	REMARKS
	Morse	Not determined					Three miles east of Latrobe on west side of large serpentine body. Active prior to 1894. (Crawford 94:118.)
126	Mount Hope	H. E. West, Box 752, Placerville	34	10N	13E	MD	Three miles north of Grizzly Flat. Active prior to 1888. A vein in granodiorite is developed by 1000-ft. adit, 200- and 100-ft. shafts. Ore treated in 10-stamp mill. (Irelan 88:178; Crawford 96:151; Tucker 19:292.)
127	Mount Pleasant	Annie S. Kirk, Box 507, Placerville	16	9N	13E	MD	(Irelan 88:178; DeGroot 90:178; Crawford 94:118; 96: 151; Tucker 19:292; Logan 22:209,301; 26:416-417; 38:241-242; herein.)
128	Nashville (Havilah, Tennessee-Nashville)	Montezuma-Apex Mining Company, c/o National Tunnel and Mines Company, 818 Kearns Building, Salt Lake City, Utah	2	8N	10E	MD	(Crawford 94:119; 96:151; Storms 00:91; Tucker 19: 292-293; Hamilton 22:30; Logan 22:4,209; 34:27; herein.)
	National						See Briarcliffe.
	New Deal						See U. S. Grant.
	New Eldorado	H. V. Colby, 210 Stone Wall Road, Berkeley 5	25	13N	9E	MD	Two and one-half miles north of Greenwood. Active around 1894. Vein containing specimen gold strikes north and dips east; developed by adit.

LODE GOLD (CONT.)

MAP NO.	CLAIM, MINE, OR GROUP	OWNER NAME, ADDRESS	SEC.	T.	R.	B & M	REMARKS
	New Eldorado (continued)						(Crawford 94:119; 96:152; Tucker 19:293.)
	New Era						See River Hill.
	No. 2 (Edmunds)	Not determined		10N	9E	MD	One mile northwest of Rescue. Active 1938, when 100 tons ore was mined. A 2-ft. vein contains high-grade gold pockets. (Logan 38:242.)
	North St. Lawrence						See St. Lawrence.
129	Oak	Arthur S. Morey et al., 91 Rico Way, San Francisco 23	32	9N	13E	MD	One mile northeast of Omo Ranch. Active 1894. A 1- to 4-ft. vein in granite containing much sulfide strikes northeast and dips 75° northwest. Developed by 400-ft. and 150-ft. drift adits. Ore treated in 5-stamp mill. (Crawford 94:119; 96:152; Tucker 19:293.)
	Oakland						See McNulty.
130	Ohio (Eagle)	Not determined	7	12N	10E	MD	One mile east of Greenwood. Active in 1894-96. A 4-ft. vein with stringers in slate strikes northwest and dips 51° northeast. Developed by 250-ft. inclined shaft. (Crawford 94:119; 96:152,157; Tucker 19:293.)
131	Ohio	Albert N. Brown et al., 5252 James Avenue, Oakland	16,21	9N	13E	MD	One mile southwest of Grizzly Flat. Active prior to 1894. A 4-ft. vein contained ore shoot several hundred ft. long. Ore averaged $12 per ton. Developed by 135-

LODE GOLD (CONT.)

MAP NO.	CLAIM, MINE, OR GROUP	OWNER NAME, ADDRESS	SEC.	T.	R.	B & M	REMARKS
	Ohio (continued)						ft. vertical shaft and and an inclined shaft. (Crawford 94:119; 96:152.)
	Old Gray						See Gray.
	Old Harmon						See Placerville Gold Mining Company.
	Old Jasper	Not determined		10N	9E	MD	Nine miles northwest of Shingle Springs. Active prior to 1896. Two parallel veins 3 and 4 ft. wide in granodiorite strike north. Developed by 200-ft. drift adit and inclined shaft. (Crawford 96:152.)
132	Omo	Not determined	32	9N	13E	MD	One mile northeast of Omo Ranch. Active 1896. A 1½-to 3-ft. vein in granodiorite strikes north, dips west near surface, but dips east in depth. Sulfides run one percent. Developed by 150-ft. adit and 64-ft. shaft. Ore treated in 125-ft. flume with pole riffles. (Crawford 96:152-153; Tucker 19:293.)
	One to Sixteen and Vulture	Not determined	6,7	10N	11E	MD	One mile north of Placerville. Long idle. (Tucker 19:293.)
133	Ophir	C. A. Clark, P.O. Box 41, El Dorado	11	9N	10E	MD	On Mother Lode two miles south of El Dorado. Active 1932-34; prospected 1951-52. A 14-inch vein and stringers in a brecciated zone in quartz porphyry and greenstone contain pyrite and arsenopyrite. Some high-grade ore produced. Developed by adit and shaft; ore treated in 2-stamp mill. (Logan 34:34-35.)

LODE GOLD (CONT.)

MAP NO.	CLAIM, MINE, OR GROUP	OWNER NAME, ADDRESS	SEC.	T.	R.	B & M	REMARKS
134	Oregon		18	10N	11E	MD	See Placerville Gold Mining Company. (Irelan 88:182-183; DeGroot 90:173; Crawford 94:119; 96:153.)
	Oregon Extension						See Placerville Gold Mining Company.
	Oregon Hill						See Placerville Gold Mining Company.
	Oriflamme						See Oro Flam.
	Oro Fino						See Big Canyon.
	Oro Fino	Not determined	33	12N	10E	MD	On Mother Lode one mile south of Garden Valley. Active 1925-26, 1930. A 4-ft. vein in amphibolite strikes northwest and dips 50° northeast. Some high-grade produced. Developed by 80-ft. inclined shaft. Ore treated at Frog Pond mill. (Logan 26:417; 34:35.)
	Oro Flam (Oriflamme)	Mill B. Maginess, et al., 1400 Ulloa Street, San Francisco	31	10N	11E	MD	On Mother Lode one mile southeast of Diamond Springs. Active 1888-90. A vein 1 to 10 ft. wide lies between slate and diorite. Developed by 350-ft. adit and 40-ft. shaft. (Irelan 88:189-190; DeGroot 90:173; Tucker 19:293.)
	Oronogo	Ruth Frazier, Garden Valley	34	12N	10E	MD	One mile south of Garden Valley on Mother Lode. Active 1953-55 with small gold output. Two parallel veins in slate with much gouge strike north and dip east. Developed by 90-ft. inclined shaft with 20-ft. drift south on 90-ft. level.

LODE GOLD (CONT.)

MAP NO.	CLAIM, MINE, OR GROUP	OWNER NAME, ADDRESS	SEC.	T.	R.	B & M	REMARKS
	Orum (Woodland)	Cecilia Simpson, 4029 Kuhnle Street, Oakland	12	9N	10E	MD	On Mother Lode three miles east of El Dorado. Active 1914. A 1-ft. vein in Mariposa slate strikes north and dips 80° east; free-milling ore with two percent pyrite. Developed by 200-ft. vertical shaft with levels at 100 and 150 ft. Ore treated in 5-stamp mill with Wilfley table (Tucker 19:300; Logan 34:35.)
135	Pacific Quartz	Placerville Gold Mining Company, Box 191, Placerville	17,18 20	10N	11E	MD	See in text under Placerville Gold Mining Company. (Irelan 88:183-186; DeGroot 90:173; Crawford 94:120; 96:163; Tucker 19:293-295; Logan 22:209; 26:417-418; 34:35-36; Bowen and Crippen 48:66,68.)
	Padre	Not determined		9N	10E	MD	On Mother Lode two miles north of Nashville. Active around 1894. A 5-ft. vein developed by 160-ft. shaft. Ore treated in 5-stamp mill. (Crawford 94:120; 96:153.)
	Paymaster	Not determined	20	13N	11E	MD	One mile south of Volcanoville. Active in 1920 and 1926. Ore treated in 10-ton Gibson mill.
	Pendelco						See Funny Bug.
	Philadelphia and Gold Note	George Allen, Sutter Creek	21,22	8N	13E	MD	Three miles southeast of Indian Diggings. Active during 1890's. A 4- to 5-ft. vein in mica schist and slate strikes north 30° west and dips 60° northeast. Ore contains pyrite and galena. An ore shoot 600 ft. in length is worked. Developed by 145- and 125-ft. shafts, and 600-ft. adit. Ore treated in 10-stamp mill. (Crawford 94:120-121; 96:153; Tucker 19:295.)

LODE GOLD (CONT.)

MAP NO.	CLAIM, MINE, OR GROUP	OWNER NAME, ADDRESS	LOCATION SEC.	T.	R.	B & M	REMARKS
	Phillips						See Big Jim.
	Pine Hill (Unity)			10N	9E	MD	Five miles northwest of Shingle Springs. Active during 1890's. A 6-ft. quartz vein in slate and greenstone strikes north contains up to 5 percent sulfides. Developed by 200-ft. shaft, drifts, and crosscuts. (Crawford 94:121; 96:153.)
	Placerville Gold Mining Company	Placerville Gold Mining Company, Box 191, Placerville	6,7 17,18 20,36	10N 11N	11E 10E	MD MD	Operations include: True Consolidated mine with the Young Harmon, Old Harmon, Halleck and Berry claims; the Van Hooker, Grass, Brown Bear, Cinnamon Bear, White Bear, and Eureka claims; Epley Consolidated mine, composed of the Epley, Faraday, Henrietta and Mammoth claims; and the Rose, Chester, Ida, Oregon and Oregon Extension claims; and Pacific. (Herein.)
	Plattsburg	Not determined	2 or 3	12N	10E	MD	Three-fourths of a mile north of Georgetown. Active in 1896. A 1- to 4-ft. vein strikes north, dips east, and contains coarse gold. Also auriferous channel on property. (Crawford 96:154.)
136	Pleasant Valley	Not determined	13	10N	12E	MD	Four miles northeast of Pleasant Valley. Active 1880's, again in 1935. A 2½- to 6-ft. vein developed by 480-ft. adit and 110-ft. shaft. (Crawford 94:154; Logan 38:244.)
137	Pocahontas	G. Q. Chase et al., 37 Lincoln Avenue, Piedmont 11	10,11 14,15	9N	10E	MD	On Mother Lode two and one-half miles south of El Dorado. Opened 1854 and intermittently active until 1896; prospected around 1939. Two veins 300 ft. apart strike

LODE GOLD (CONT.)

MAP NO.	CLAIM, MINE, OR GROUP	OWNER NAME, ADDRESS	SEC.	T.	R.	B & M	REMARKS
	Pocahontas (continued)						northwest and dip northeast, in diorite porphyry. Ore ranged from $4 to $25 per ton in value. Developed by 400- and 300-ft. inclined shafts and about 1700 ft. of drifts. Ore was treated in 10-stamp mill. (Crawford 94:121; 96:154; Storms 00:95; Tucker 19:295; Logan 34:36.)
	Polar Bear (Empire Group, White Bear)	Martha Agnes Weber, Placerville and Maryland Casualty Company, 240 Sansome Street, San Francisco	29,32	9N	13E	MD	Three miles south of Grizzly Flat. Long idle. (Tucker 19:295.)
	Poor	F. L. Veerkamp, c/o Stiner's Station, Kelsey	12	11N	10E	MD	One mile northwest of Kelsey. Active around 1938 when 200 to 300 tons ore were mined. Quartz stringers in amphibolite strike north 15 west and dip steeply northeast. Developed by open cuts and 37-ft. shaft. (Logan 38:244.)
	Potosi	Not determined	33	9N	13E	MD	One mile east of Omo Ranch. Active in 1908.
	Potter						See Rising Sun.
	Poverty Point						See Guildford.
138	Pyramid (Gold Reserve)	Rhoads Grimshaw, Auburn	12,13	10N	9E	MD	(Crawford 94:121; 96:154; Tucker 19:295; Logan 35: 23; 38:244-246; herein.)
	Rainbow (Wild West)	Not determined	21	12N	10E	MD	One and one-half mile northwest of Garden Valley. Active 1896. A vein system as much as 23 ft. wide strikes north and dips 45° east with porphyry hanging wall and slate footwall. Developed by open cuts and 25-ft. shaft.

LODE GOLD (CONT.)

| MAP NO | CLAIM, MINE, OR GROUP | OWNER NAME, ADDRESS | LOCATION | | | | REMARKS |
			SEC.	T.	R.	B & M	
	Rainbow (continued)						Ore treated in 4-stamp mill. (Crawford 96:155; Tucker 19:295.)
	Rattler	Not determined	20	10N	11E	MD	Two miles south of Placerville by Weber Creek. Active prior to 1894. Developed by two adits. (Crawford 94: 122; 96:155.)
	Red Hill	Not determined		12N	10E	MD	Two miles northwest of Garden Valley. Active around 1914. A northwest-striking vein dips northeast and occurs in slate. Developed by 100-ft. inclined shaft and 350-ft. drift. Ore treated in 2-stamp mill. (Tucker 19:295-296.)
	Red Rover	Not determined		9N	10E	MD	Three miles southeast of El Dorado. Active around 1894 and again in 1920. A 6-in. to 3-ft. vein strikes north and dips 60° east with granite hanging wall and slate footwall. Developed by 130- and 30-ft. inclined shafts and 115 ft. of drifts. (Crawford 94:122; 96:155; Logan 20:427-428.)
	Red Top						See Red Wing.
139	Red Wing (Red Top)	David Loofbourrow, 333 California Street, San Francisco	12,13	9N	10E	MD	Three miles south of El Dorado on Mother Lode. Active around 1914-22 and 1926. A 5-ft. vein in Mariposa slate strikes north 10° east and dips 70° southeast. Developed by upper 125-ft. adit and lower 525-ft. crosscut adit and drifts. Ore treated in a 2-stamp mill. (Tucker 19:296; Logan 23a:301.)

LODE GOLD (CONT.)

MAP NO.	CLAIM, MINE, OR GROUP	OWNER NAME, ADDRESS	LOCATION SEC.	LOCATION T.	LOCATION R.	LOCATION B & M	REMARKS
	Ribbon Rock						See Miller.
140	Richelieu	Mary Ann Simpson, c/o J. P. Conroy, Box 212, El Dorado	12	9N	10E	MD	On Mother Lode three miles south of El Dorado. Active 1932. Small amounts of gold in quartz stringers in slate and gouge. Developed by shaft and 275-ft. adit. Small amount of ore treated at Church mill. (Logan 38:246-247.)
	Richmond	L. G. Fancher, Box 684, Jackson	4 33	8N 9N	13E 13E	MD MD	One mile south of Fairplay. Long idle. Ore treated in 8-stamp mill. (Tucker 19:296.)
	Rising Sun (Potter)	Oviedo Le Bourreau, Box 592, Placerville	14	11N	10E	MD	One mile northwest of Kelsey. Long idle. (Tucker 19:296.)
141	River Hill group (Bell, Gentle Annie, Hall Consolidated, Lucky Star, Lyon, New Era).	A. A. McKinnon, Placerville	6 36	10N 11N	11E 11E	MD MD	A consolidation of claims on Mother Lode one and one-half miles northwest of Placerville. Originally worked about 1865. Active 1890 to 1906; large total output. Five parallel veins in Mariposa slate strike northwest and dip northeast; ore shoots were up to 30 ft. in width and 150 ft. in length; yielded $2.50 to $4.80 per ton in early 1900's. Developed by 600- and 1550-ft. inclined shafts, 2400-ft. adit, and much drifting. Ore treated originally in 10-stamp mill; about 1901, a 20-stamp mill was erected. (DeGroot 90:177; Crawford 94:111; 96:141-142; Storms 00:94; Tucker 19:296; Logan 34:37.)
	Rocky Bar	Not determined		12N	10E	MD	One-half mile east of Greenwood. Active prior to 1894. A 1-ft. vein in slate strikes northwest. Developed by open cuts and shallow shafts. (Crawford 94:122.)

LODE GOLD (CONT.)

MAP NO.	CLAIM, MINE, OR GROUP	OWNER NAME, ADDRESS	LOCATION SEC.	T.	R.	B & M	REMARKS
	Roscoe	Not determined					Three miles northeast of Latrobe. Prospected in 1896. A 12-ft. vein strikes north and dips west. (Crawford 96:155.)
	Rose		18	10N	11E	MD	See Placerville Gold Mining Company. (Irelan 88:182-183; Crawford 94:122; 96:155.)
	Rosecrans						See Rosecranz.
142	Rosecranz (Rosecrans)	Arthur S. Morey, 91 Rico Way, San Francisco	21	12N	10E	MD	(Irelan 88:171; DeGroot 90:176; Crawford 96:155; Tucker 19:296; Logan 38:248-249; herein.)
143	Rose Kimberly	Hattie M. Grien, Rescue	10,11	10N	9E	MD	Two miles northwest of Rescue. Active 1895 and again in 1938. Two lensoid veins in gabbro contain pyrite, galena, chalcopyrite; assays range from $1 to $5 per ton. Developed by 220-ft. inclined shaft with levels at 60, 120, and 220 ft. (Logan 38:247-248.)
	Ruby Consolidated	Not determined	17,18 19	13N	11E	MD	One mile south of Volcanoville. Intermittently active 1928-40. Gold associated with serpentine. Developed by 900-ft. adit and 180-ft. vertical shaft. Ore treated in 2-stamp mill. Also a chrome mine; see Chromite tabulated list.
	Ryan	Not determined	24	11N	10E	MD	One mile south of Kelsey. Long idle. (Tucker 19:296.)
	Safeguard No. 1						See Bret Harte.

LODE GOLD (CONT.)

MAP NO.	CLAIM, MINE, OR GROUP	OWNER NAME, ADDRESS	LOCATION SEC.	T.	R.	B & M	REMARKS
144	St. Clair	J. Peters, Kelsey	14	11N	10E	MD	One mile northwest of Kelsey. Active prior to 1915 and again around 1940. (Tucker 19:298.)
145	St. Lawrence	St. Lawrence Gold Mining Company, c/o Wells Fargo, 4 Montgomery Street, San Francisco	3	11N	10E	MD	On Mother Lode one and one-half miles southeast of Garden Valley. Active 1867-78 when $465,000 was produced. A 6-ft. vein in Mariposa slate strikes north 10° west and dips 60° northeast. Ore ranged from $8.54 to $27 per ton in value, but average was $10 to $17. Developed by 900-ft. inclined shaft with levels at 100 ft. and 200-ft. winze sunk from 900-ft. level. Ore treated in 20-stamp mill. (Preston 92:202; Crawford 96:156; Tucker 19:298; Logan 34:40-41.)
	St. Lawrence						See Stillwagon.
	Salisbury	Not determined	33	10N	11E	MD	Three miles southeast of Diamond Springs. Active around 1896. Vein strikes northeast and dips 70° southeast. Developed by 40-ft. inclined shaft. Ore treated in 40-ton mill equipped with two Huntington mills. (Crawford 96:156.)
	Schleifer	Not determined					On east side of Big Canyon seven miles south of Shingle Springs. Active prior to 1894. A low-grade body of auriferous pyrite. (Crawford 94:122.)
	Schneider						See Mathenas Creek.

LODE GOLD (CONT.)

| MAP NO. | CLAIM, MINE, OR GROUP | OWNER NAME, ADDRESS | LOCATION | | | | REMARKS |
			SEC.	T.	R.	B & M	
	Selby	Not determined	29	10N	11E	MD	One mile east of Diamond Springs. Active prior to 1900. Developed by 240-ft. shaft. (Storms 00:93; Tucker 19:296.)
	Shan Tsz						See Shaw.
	Sharp	Not determined		10N	11E	MD	Six miles east of Placerville. Opened about 1870; again active in 1890. A 12-ft. vein strikes northwest and dips northeast with granite footwall and slate hanging wall. Developed by 54-ft. shaft and 110- and 108-ft. adits. Ore treated in 10-stamp mill. (Irelan 88:194; Crawford 94;123; 96:156.)
146	Shaw (Shan Tsz, Volo)	Evelyn R. Purser, c/o Richard P. Lyons, 110 Bank Building, South San Francisco Leased by Volo Mining Company, 464 Main Street, Placerville	21	10N	10E	MD	(Irelan 88:193; DeGroot 90:181; Crawford 94:481; Tucker 19:296-297; herein.)
	Shepard						See Lucky Marion.
147	Sherman	Sherman Mining and Milling Company, Placerville	8	10N	11E	MD	On Mother Lode one mile north of Placerville. Active 1905, 1908-11 with production of $136,000. A 5-ft. vein in Mariposa slate strikes north.20° west and dips 74° northeast. Ore was free milling with 3 percent

LODE GOLD (CONT.)

MAP NO.	CLAIM, MINE, OR GROUP	OWNER NAME, ADDRESS	SEC.	T.	R.	B & M	REMARKS
	Sherman (continued)						pyrite. Ore yielded $3.15 to $6.50 per ton. Developed by 750-ft. inclined shaft with levels at 100, 200, 300, 400, 500, and 750 ft; 350-ft. winze sunk from 750-ft. level; 5500 ft. of drifts. Ore treated in 10-stamp mill with tables and vanners. (Tucker 19:297-298; Logan 34:38.)
148	Shumway	Wilber E. Timm, et al., c/o County Courthouse, Placerville	7	11N	11E	MD	One mile southeast of Spanish Flat. Prospected in 1938. Developed by 100-ft. shaft and 300-ft. crosscut adit. (Logan 38:249.)
	Sir Walter Raleigh						See Elliott.
149	Skipper (Esperanza)	H. R. Swartz, Greenwood	7	12N	10E	MD	(Crawford 96:140; Tucker 19:285; herein.)
	Slate Mountain	E. V. Little, Garden Valley	35	12N	11E	MD	Three miles northwest of Slate Mountain. Active intermittently 1921-41; 1951. A 1½- to 8-ft. vein developed by 600-ft. crosscut adit and 2400 feet of drifts. Ore treated in 10-stamp mill.
150	Sliger	Sliger Gold Mining Company, 3628 Fulton Street, San Francisco	36	13N	9E	MD	(Crawford 94:123; 96:157; Tucker 19:297; Logan 23:157; Tucker 19:297; Logan 23:143; 26:418; 34:38-40; 38:250; herein.)
	South Ohio	Not determined		12N	10E	MD	One and one-half miles east of Greenwood. Active in 1896. An east-striking and north-dipping vein in slate developed by 100-ft. adit. (Crawford 96:157.)
	Spanish Dry Diggings						See Grit.

LODE GOLD (CONT.)

MAP NO.	CLAIM, MINE, OR GROUP	OWNER NAME, ADDRESS	LOCATION SEC.	T.	R.	B & M	REMARKS
	Springfield						See Union.
	Standard	John Fitzgerald, 1028 Pendegast Street, Woodland	24	11N	10E	MD	Quarter of a mile north of Coloma. Active around 1894. Seems 2 to 12 in. wide occur near granite-slate contact. Developed by 230-ft. adit. (Crawford 94:123; 96:157.)
151	Starlight	Starlight Mining Co., 2626 Baker Street, San Francisco	3,10	9N	10E	MD	On Logtown Ridge two and one-half miles south of El Dorado. Active 1894-96. A 3-ft. vein strikes northwest and dips 40° northeast. Developed by 500-ft. vertical and two shallow inclined shafts and drifts and cross-cuts. Ore was treated in 10-stamp mill. (Crawford 94: 123; 96:157; Tucker 19:298.)
152	Stillwagon (St. Lawrence)	Not determined	32,33	9N	13E	MD	One mile northeast of Omo Ranch. Active 1890-94 and 1908-14. A 2- to 3-ft. vein in granite strikes north-east and dips southeast and is rich in sulfides. De-veloped by 400- and 200-ft. adits. Ore treated in 5-stamp mill. (DeGroot 90:178; Crawford 94:123-124; Tucker 19:298.)
	Stony Point						See Adams Gulch.
153	Stuckslager	Beryl E. McKenney, 31 Lake Vista Avenue, Daly City 25	24	11N	9E	MD	(Hanks 86:43; DeGroot 90:178; Crawford 94:124; 96:158; Tucker 19:298; herein.)
	Studhorse	Not determined		12N	9E	MD	Two miles west of Greenwood. Active around 1896. An east-striking and north-dipping vein in slate developed by 30- and 40-ft. shafts. (Crawford 96:158.)

LODE GOLD (CONT.)

MAP NO.	CLAIM, MINE, OR GROUP	OWNER NAME, ADDRESS	LOCATION SEC.	LOCATION T.	LOCATION R.	LOCATION B & M	REMARKS
154	Sugar Loaf	Sugar Loaf Mining Company, Route 1, Box 57, Shingle Springs	1,12	8N	9E	MD	(Crawford 94:124; 96:158; herein.)
	Sullivan						See Adams Gulch.
	Sunday	Not determined	16	9N	13E	MD	One and one-half miles west of Grizzly Flat. Active around 1894. A 1- to 3-ft. vein in syenite strikes north. Developed by 110-ft. shaft, 80-ft. drift, 300-ft. adit, and open cuts. (Crawford 94:124; 96:158; Tucker 19:298.)
155	Sunrise	John Fitzgerald, 1028 Pendegast Street, Woodland	24	11N	10E	MD	On Mother Lode one mile southeast of Kelsey. Active around 1896. Three parallel veins 9, 15, and 30 ft. thick in slate with associated talc and calcite strike north and dip east. Developed by 320-ft. crosscut adit and several shallow shafts, ore treated in 10-stamp mill. (Crawford 96:158-159; Tucker 19:299; Logan 34:41.)
156	Superior (Tin Cup)	Henry A. Alker and Elliott S. McCurdy, Trustees for Estate of Emma H. Rose, 220 Montgomery Street, San Francisco	20,29 30	10N	11E	MD	On Mother Lode one mile east of Diamond Springs. Located in 1867, active around 1888-90, and again in 1900. Three parallel veins strike north and dip 45° east with slate footwall and greenstone hanging wall. Ore yielded $4.80 to $15 per ton. Developed by 250- and 700-ft. adits and 400-ft. inclined shaft. Ore treated in 10-stamp mill. (Irelan 88:187-189; DeGroot 90:172; Crawford 94:124; 96:159; Storms 00:94; Tucker 19:299; Logan 34:41.)

LODE GOLD (CONT.)

MAP NO.	CLAIM, MINE, OR GROUP	OWNER NAME, ADDRESS	SEC.	T.	R.	B & M	REMARKS
	Syracuse	Not determined	33	9N	13E	MD	One mile east of Omo Ranch. Active in 1908.
	Tapioca						See California Consolidated.
157	Taylor (Idlewild)	W. N. Guth, 2024 N. New England Avenue, Chicago, Illinois	21,30	12N	10E	MD	(Irelan 88:168-171; DeGroot 90:176; Preston 92:205; Crawford 94:113; 96:145; Tucker 19:299; Logan 22:209,210; 34:41-42; herein.)
	Tennessee-Nashville						See Nashville.
	Threlkel (Winton)	United States of America (withdrawn for Federal reservoir site)	16	11N	8E	MD	Immediately east of the Zantgraf mine one mile south of Rattlesnake Bridge. Active 1924-26 and 1937. Several thin veins of good grade in granodiorite. Developed by adit. Ore treated in 2-stamp mill. (Logan 26:418; 38:255.)
	Tin Cup						See Superior.
	Treat	Helen B. Treat, 93 San Carlos Avenue, Sausalito	4	9N	13E	MD	Two and one-half miles north of Grizzly Flat. Active prior to 1888 and again in 1896. Vein in granodiorite strikes north and dips west. Developed by 100-ft. vertical shaft and adits. (Irelan 88:178; Crawford 96:159; Tucker 19:299.)
158	Trench (Yellowjacket)	Not determined	18	13N	11E	MD	In Quartz Canyon one mile south of Volcanoville. Active prior to 1894. A vein strikes north. (Crawford 94:125; 96:159; Tucker 19:299.)

LODE GOLD (CONT.)

MAP NO.	CLAIM, MINE, OR GROUP	OWNER NAME, ADDRESS	SEC.	T.	R.	B & M	REMARKS
	True Consolidated		6,7	10N	11E	MD	See Placerville Gold Mining Company. (Irelan 88:180-181; Crawford 94:125; 96:144.)
159	Tullis (Diamond)	Starlight Mining Co., 2626 Baker Street, San Francisco	1	9N	10E	MD	One mile southeast of El Dorado. Active 1896. A $2\frac{1}{2}$-ft. vein in quartz porphyry strikes north and dips 30° east. Developed by 200-ft. inclined shaft and 175 ft. of drifts. (Crawford 96:159-160; Tucker 19:299.)
160	Union (Springfield)	El Dorado Mining Company, P.O. Box 1724, Spokane 7, Washington	12	9N	10E	MD	(Hanks 86:43; Irelan 88:167; Storms 00:92; Tucker 19:299; Logan 22:209-210; 26:418-419; 34:42-43; 38:251-253; herein.)
	Union Consolidated						See Alpine.
	Unity						See Pine Hill.
	U. S. Grant (New Deal)	Not determined	32	11N	12E	MD	North of Mt. Danaher. Active during 1870's when ore treated in 10-stamp mill; prospected in 1936. A 1- to 4-ft. vein contained two ore shoots. Developed by shaft and 100-ft. crosscut adit with drifts 400 ft. north and 200 ft. south. (Logan 38:253.)
	Utah Apex						See Montezuma.
161	Valdora	Not determined	16	9N	13E	MD	Adjoining Mt. Pleasant mine on the south, one-half mile west of Grizzly Flat. Active around 1888. Granite wall rock. Developed by 110-ft. vertical shaft.

LODE GOLD (CONT.)

MAP NO.	CLAIM, MINE, OR GROUP	OWNER NAME, ADDRESS	SEC.	T.	R.	B & M	REMARKS
	Valdora (continued)						(Irelan 88:178; Crawford 96:161.)
	Van (Vann)	V. M. Fishback, Box 164, Georgetown	2	12N	10E	MD	Three-quarters of a mile north of Georgetown. Long idle. (Crawford 94:126; 96:161; Tucker 19:300.)
162	Vandalia	G. L. Tripp, Route 1, Box 5, Shingle Springs	19	9N	10E	MD	(Irelan 88:172-173; DeGroot 90:178; Crawford 94:126; Storms 00:96-98; Logan 22:301; 26:419; 38:254; herein.)
163	Vandergreft	Karl Skuja, El Dorado	26	9N	10E	MD	Three miles north of Nashville. Active prior to 1914. Developed by 250-ft. inclined shaft and 100-ft. adit. Ore treated in 10-stamp mill. (Tucker 19:300.)
164	Van Hooker		6 / 36	10N / 11N	11E / 10E	MD / MD	See Placerville Gold Mining Company. (Irelan 88:181-182; DeGroot 90:173; Crawford 96:161.)
	Vann						See Van.
165	Veerkamp (Gold Coin)	Ray A. Veerkamp, Garden Valley	33	12N	10E	MD	(Logan 34:37-38; 38:247; herein.)
	Victoria	Not determined	4	10N	9E	MD	Four miles northwest of Rescue near Boulder mine. Active 1924-26. Vein developed by 30-ft. shaft and 50-ft. drift. Ore which yielded $8 per ton, was treated in 2-stamp mill. (Logan 24:178; 26:419.)
	Volo						See Shaw; also see Volo mill in Copper section.

LODE GOLD (CONT.)

MAP NO.	CLAIM, MINE, OR GROUP	OWNER NAME, ADDRESS	LOCATION				REMARKS
			SEC.	T.	R.	B & M	
	Wagner	Camilla D. Heald, 624 South 13th Street, San Jose	11	8N	10E	MD	Near Lotus; prospected in 1925. (Logan 26:419.)
166	War Eagle	H. J. Picchetti, 1115 Fairview Avenue, San Jose	24	11N	10E	MD	On Mother Lode one mile southeast of Kelsey. Ore occurs in pockets in 18-in. vein. Developed by 80-ft. shaft and 150-ft. adit. (Logan 34:43.)
	Webber						See Placerville Gold Mining Company.
	Webster	Georgetown Lumber Company, Georgetown	31	13N	11E	MD	In Quartz Canyon two miles south of Volcanoville. Active around 1894. A 1-ft. vein between slate and serpentine strikes north 15° east and dips 55° southeast. Developed by 200- and 300-ft. crosscut adits. (Crawford 94:126; 96:161; Tucker 19:300.)
167	Welch	Dollie K. Goss and H. S. McDaniel, 315 Park View Terrace, Oakland	7	12N	10E	MD	Half a mile northeast of Greenwood. Active 1894-96. A 3- to 8-ft. vein in slate. Ground was sluiced and then vein was developed by 100-ft. inclined shaft and 150-ft. crosscut adit. (Crawford 94:126; 96:161; Tucker 19:300.)
	White Bear						See Polar Bear.
	White Bear		36	11N	10E	MD	See Placerville Gold Mining Company. (Irelan 88:182; Crawford 96:161.)

LODE GOLD (CONT.)

| MAP NO. | CLAIM, MINE, OR GROUP | OWNER NAME, ADDRESS | LOCATION | | | | REMARKS |
			SEC.	T.	R.	B & M	
168	White Owl	Kate E. Hennesey, 515 Park Avenue, New York, N.Y.	23	11N	11E	MD	Near Red Bird Creek two miles southeast of Mosquito Camp. Active 1938. A 1½- to 3-ft. vein in granodiorite strikes northeast; ore yielded up to $65 per ton. Developed by 65-ft. inclined shaft. Ore treated in 2-ton Gibson mill. (Logan 38:254-255.)
	Wiedebush	Not determined	20	13N	11E	MD	Two miles south of Volcanoville. Active 1920-26. A 70-ft. ore shoot 3 to 4 ft. wide developed by adit. Ore treated in small roller mill. (Logan 21:428; 26:419.)
	Wild Cat	Not determined	20,21,28,29	12N	10E	MD	Two miles northwest of Garden Valley. Active in 1926.
	Wild West						See Rainbow.
	Wilhelm and Last Chance	W. J. White, Cool	25	12N	8E	MD	Four miles southeast of Auburn. Long idle. (Tucker 19:300.)
	Wiltshire	Elizabeth M. Green, c/o W. A. Green, 205 Courthouse, Sacramento	36	9N	10E	MD	A "pocket" mine near Nashville. Active around 1926. (Logan 26:419.)
	Winton						See Threlkel.
	Woodland						See Orum.
	Woodside - Eureka						See Beebe. (Preston 92:200; Crawford 94:126; 96:161; Tucker 19:300.)

LODE GOLD (CONT.)

MAP NO.	CLAIM, MINE, OR GROUP	OWNER NAME, ADDRESS	LOCATION				REMARKS
			SEC.	T.	R.	B & M	
	Yellowjacket						See Trench.
	Young Harmon		7	10N	11E	MD	See Placerville Gold Mining Company.
169	Zantgraf (Montauk Consolidated, Zentgraf)	United States Government (land withdrawn for Federal reservoir)	16	11N	8E	MD	(Irelan 88:200-202; DeGroot 90:178; Crawford 96:161-162; Tucker 19:300; Logan 22:209-210; 24:8; 26:419-420; 38:255-257; herein.)
	Zentgraf						See Zantgraf.

PLACER GOLD

MAP NO.	CLAIM, MINE, OR GROUP	OWNER NAME, ADDRESS	SEC.	LOCATION T.	R.	B & M	REMARKS
	Allen dredge	Allen Gold Dredge Company, Lotus					Operated suction dredge on Bacchi Ranch near Lotus during 1945-47.
	Alveoro	Not determined	3,10	10N	11E	MD	Drift mine one-half mile northeast of Smith's Flat. Active around 1908. Channel deposit 100 to 300 ft. wide and 6 to 30 ft. deep; gravel is partially cemented and capped by andesite. Developed by 4000-ft. adit and 400- and 500-ft. inclined shafts. (Tucker 19:300-301.)
	Amelia		9	13N	11E	MD	Two miles east of Volcanoville. Developed by 600-ft. adit. See also Two Channel Mining Company which worked property many years ago. (Logan 38:251.)
	Armstrong and Roberts	Not determined	29	9N	13E	MD	Drift mine at Henry Diggings three miles south of Grizzly Flat. Active 1894. A northeast-striking channel 60 ft. wide and 5 ft. thick is developed by 600-ft. adit. (Crawford 94:101-102; 96:132; Tucker 19:301.)
	Avansino	Alida L. Avansino, Route 3, Box 196, Placerville	29	10N	12E	MD	Drift mine near Pleasant Valley. Active around 1893; prospected in early 1930's. Channel and bench gravels developed by 107-ft. shaft with 57-ft. north drift on 90-ft. level and 307-ft. south drift on 107-ft. level. (Logan 38:216-217.)
170	Badger Hill	G. W. McMillen, 1342 Gilman Street, Berkeley	28	11N	12E	MD	At Badger Hill seven miles east of Placerville on Tertiary channel of South Fork of American River. Some rich gravel interbedded with rhyolite tuff mined by drifting and sluicing. (Lindgren 11:170; Tucker 19:301.)

PLACER GOLD (CONT.)

MAP NO.	CLAIM, MINE, OR GROUP	OWNER NAME, ADDRESS	SEC.	T.	R.	B & M	REMARKS
	Ball	Not determined		8N	13E	MD	Drift mine three and one-half miles southeast of Omo Ranch. Active around 1935. Well-cemented gravel 80-ft. wide. Developed by 1,250-ft. adit to channel and 600-ft. drift south. (Logan 38:217.)
	Barker dredge	Barker Corporation, Patterson					Operated dragline dredge on the Explorers property in 1942. (Averill 46:255.)
	Barnes						See Channel Bend.
171	Bella Vista	Harold C. Welch et al., P.O. Box 146, Sutter Creek	8	8N	12E	MD	Drift mine three miles northeast of Aukum. Active 1936. Two channels, one above other, trend northeast, lie on granodiorite and capped by rhyolite. Contained $1 to $1.50 per yard. Developed by 400-ft. drift adit, raises, and 200-ft. drift. Gravel treated in washing plant and sluice. (Logan 38:218-219.)
	Bend						See Channel Bend.
	Benfeldt (Rogers)	Not determined	10	10N	11E	MD	Drift mine at Smith's Flat. Active 1888-96 and 1916-19. Channel trends north, averages 5 ft. in thickness and is 50 to 120 ft. wide; slate bedrock; gravel yielded $2 to $8 per ton. Developed by 750-ft. inclined shaft and drifts. Gravel treated in 10-stamp mill and 150-ft. sluice. (Irelan 88:197-198; DeGroot 90:179; Crawford 96:133; Lindgren 11:177; Tucker 19:301.)
	Big Canyon dredge	Big Canyon Dredging Company, Fresno					Operated 3-cu. yd. dragline dredge on Big Canyon and Deer Creeks from 1937 to 1942. (Logan 38:363; Averill 46:255.)

PLACER GOLD (CONT.)

MAP NO.	CLAIM, MINE, OR GROUP	OWNER NAME, ADDRESS	LOCATION SEC.	T.	R.	B & M	REMARKS
	Bitters						See Two Channel Mining Company.
	Black Gold	Alida L. Avansino, Route 3, Box 196, Placerville	29	10N	12E	MD	Drift mine in Pleasant Valley district adjoining Hinds drift mine. Active 1930-31 and 1936, with an output of several thousand dollars. Bench of fine loose gravel developed by 60-ft. shaft with drifts 100 ft. west, 280 ft. north, and 127 ft. east. Gravel treated in two 10-stamp mills. (Logan 38:223.)
	Blacklock	Not determined	8,17	10N	11E	MD	Drift and hydraulic mine one mile northeast of center of Placerville. Active 1896. A 60-ft. bank with 4-ft. thick channel hydraulicked and later worked by drifting. (Crawford 96:134; Tucker 19:301.)
	Blair	E. W. Meyers, Box 336, Camino	10,11	10N	12E	MD	Two miles southeast of Camino. Prospected and drilled in 1890. Channel is 300 to 400 ft. wide. (DeGroot 90: 179; Crawford 94:104.)
	Blakeley						See Maple Leaf.
	Boles	United States of America (withdrawn for Federal reservoir site)					On American River three miles upstream from Rattlesnake Bridge. In 1924-25, gravels from bed of river excavated by barge-mounted suction pump which discharged to sluice boxes. Diver directed the nozzle. (Logan 26:438.)
	Bottle Hill	Not determined	27	13N	10E	MD	Two miles northwest of Georgetown. Isolated patch of auriferous gravel capped by andesite mined by drifting many years ago. (Lindgren 11:169.)

PLACER GOLD (CONT.)

MAP NO.	CLAIM, MINE, OR GROUP	OWNER NAME, ADDRESS	SEC.	T.	R.	B & M	REMARKS
	Boulder	Pilot Hill Mining Co., Pilot Hill	6	11N	9E	MD	At Pilot Hill. Old channel remnant 20 to 40 ft. deep worked by power shovel, and gravel treated in stationary washing plant in 1936. Yield was 13 to 60 cents per yard. (Logan 38:243.)
	Brass						See Murzo.
	Bronson dredge	W. B. Bronson, Georgetown					Operated dragline and non-floating washing plant on Browning Ranch in 1947.
172	Buckeye Hill (Flora)	W. E. Colby, 2416 East Flora, Stockton	12	13N	10E	MD	On Buckeye Point two miles west of Volcanoville. Active 1890's and early 1930's. Channel trends northeast; is as much as 1,000 ft. wide with alternating layers of gravel and cemented material. Mined by hydraulicking and drifting. Developed by 400-ft. bedrock adit. Gravel yielded $1.33 per ton. (Crawford 94:105; 96:136.)
	Buck's Bar	Not determined	6	9N	12E	MD	Northeast of Buck's Bar on North Fork of Cosumnes River. A gravel deposit 8 to 16 ft. deep worked by dragline in 1936. (Logan 38:227.)
	Burt Alley	O.S. Strawn, Box 255, Georgetown	29,30	13N	11E	MD	Three miles south of Volcanoville. Gravel deposit worked around 1894. (Crawford 94:105; 96:136.)
	California Mohawk						See Fairplay.
	Canyon Creek						See Gold Bug. (Crawford 94:105; Tucker 19:301.)

PLACER GOLD (CONT.)

MAP NO.	CLAIM, MINE, OR GROUP	OWNER NAME, ADDRESS	SEC.	T.	R.	B & M	REMARKS
	Carpender	Not determined					Worked in conjunction with Kumfa drift mine 1911, 1913, 1919, 1928, and 1936; see also Kumfa.
	Carrie Hale	Not determined	34	9N	13E	MD	Drift mine at Henry's Diggings, three miles south of Grizzly Flat. Active around 1894. Channel extends northeast, is 60 ft. wide, and up to 5 ft. thick. Developed by 400-ft. bedrock adit; mined in 12-ft. breasts. Pay streak in blue gravel on granite bedrock. (Crawford 94:105-106; 96:136.)
	Cedar Ravine	Not determined	17	10N	11E	MD	In Cedar Ravine one mile south of Placerville. Active prior to 1890. Well-cemented gravel was mined and treated in 10-stamp mill. (DeGroot 90:180.)
173	Cedar Spring	Not determined	17	10N	11E	MD	Drift mine at Cedar Ravine one mile south of Placerville on Green Mountain channel. Active 1870's to around 1901. Channel is crooked, 300 to 600 ft. wide, and pay gravel is 4 to 6 ft. thick; some benches. A deep channel with benches also mined. Developed by 900-ft. south-southeast adit and 75-ft. incline. Also a lower aidt. (Lindgren 11:175-176.)
	Cement Hill	Not determined	27,28 33,34	13N	10E	MD	Drift mine on Cement Hill three miles northwest of Georgetown. Active 1894-96. An east-west channel prospected by 750- and 600- ft. adits. (Crawford 94: 106; 96:137.)

PLACER GOLD (CONT.)

MAP NO.	CLAIM, MINE, OR GROUP	OWNER NAME, ADDRESS	SEC.	T.	R.	B & M	REMARKS
174	Channel Bend (Barnes, Bend, Gray Eagle Bar, McCall)	M. T. Duffey, 6009 Santa Fe Avenue, Huntington Park	5	13N	11E	MD	Drift mine two miles northeast of Volcanoville by Middle Fork of the American River. A bend in a channel of the Tertiary Middle Fork worked through 136-ft. shaft and 200- and 300-ft. drifts during the 1890's. (Crawford 96:137; Tucker 19:301.)
	Chili Ravine	Not determined	19	10N	11E	MD	Drift mine in Chili Ravine two miles south of Placerville. Active 1870-90 and 1912-15. Channel trends north; gravel is well cemented and 3 to 12 ft. thick. Developed by a 1200-ft. crosscut adit, 700-ft. drift adit, and air raises. (Irelan 88:194-196; DeGroot 90:180; Crawford 94:106.)
	Christian	Not determined					Drift mine at Henry Diggings three miles south of Grizzly Flat. Worked intermittently since 1949 by I. H. Campion, Somerset. See also Slide and Payne drift mines.
	Claghorn (Growers)	Not determined		8N	12E	MD	Drift mine on Cedar Creek two miles south of Fairplay. Active around 1896. A northeast-trending Tertiary channel with granite bedrock developed by 200-ft. adit. (Crawford 96:138.)
	Clark	Not determined	16	10N	11E	MD	Drift mine on south side Texas Hill, one mile southeast of Placerville. Active during early days of gold rush. Several benches on tributary of Tertiary South Fork of American River developed by north-trending adit several hundred feet long. (Lindgren 11:178-179.)
	Clarksville dredge	F. O. Bohnett, Santa Barbara					Operated dry land dredge near Clarksville in 1939.

PLACER GOLD (CONT.)

| MAP NO. | CLAIM, MINE, OR GROUP | OWNER NAME, ADDRESS | LOCATION | | | | REMARKS |
			SEC.	T.	R.	B & M	
	Confederate	Not determined	8	8N	12E	MD	Drift mine two and one-half miles southwest of Fairplay. Active 1896. Two adits, 250 and 200 ft. in length driven in Tertiary channel on granite bedrock. (Crawford 96: 138; Tucker 19:301.)
	Cooley	Not determined	17	13N	11E	MD	Drift mine at Volcanoville. Prospected by Wm. Ogle in 1934 and 1936; some gold output.
175	Coon Hollow (includes Excelsior claim)	Not determined	18,19	10N	11E	MD	Drift and hydraulic mining one mile south of Placerville. Drifted from 1852 to 1861 and hydraulicked from 1861 to 1871. Gold output from this area totaled $10,000,000. Gravel yielded about $1 per cubic yard. Accumulated tailings being mined for aggregate; see Stone section. See also Excelsior. (Whitney 80:119; Lindgren 11:172, 174.)
	Danaher						See Roundabout.
	Diamond Springs		24,25	10N	11E	MD	At Diamond Springs. Extensive deposit of Tertiary gravel; portion mined by hydraulicking many years ago. (Lindgren 11:170.)
	Dividend	Not determined	3	10N	9E	MD	On Pichem Creek four miles northwest of Rescue. Active 1880's, 1890's and 1912-15. Extensive deposit of gravel 1 to 3 ft. thick on granite bedrock worked by ground sluicing. (Crawford 94:108; 96:139; Tucker 19:301.)
	Dorsey	Not determined	17	8N	13E	MD	One mile northeast of Indian Diggings. Long idle.
	Eagle	Not determined	31	9N	13E	MD	One mile northwest of Omo Ranch. Long idle.

PLACER GOLD (CONT.)

MAP NO.	CLAIM, MINE, OR GROUP	OWNER NAME, ADDRESS	LOCATION SEC.	T.	R.	B & M	REMARKS
	El Dorado Dredge	El Dorado Dredging Corporation, Greenwood					Operated 1½-cu. yd. dragline dredge on Greenwood, Coloma, Rock Canyon, and Irish Creeks, 1940-42 and 1948. (Averill 46:255.)
	El Dorado Water and Deep Gravel Mining Company						Operated hydraulic and drift mines in the Placerville area including the Excelsior. Held claims in Coon Hollow. See Coon Hollow and Excelsior.
	Excelsior	Placerville Gold Mining Company, Box 191, Placerville	18,19	10N	11E	MD	At Coon Hollow one mile south of Placerville. About $5,000,000 recovered from this claim by drifting and hydraulicking between 1852 and 1871. Mined by drifting during 1907-11. Gravel was treated in 10-stamp mill. Tributary of Tertiary South Fork of American River contained three pay streaks. See also Coon Hollow. (Whitney 80:98,114,119; Lindgren 11:172,174-175.)
	Fairplay (California Mohawk)	Not determined	34	9N	12E	MD	Drift mine just east of Fairplay. Long idle. At one time this property was owned by the California Mohawk Mining Company. (Tucker 19:301.)
	Ferriera		29	10N	12E	MD	One mile south of Newtown. Prospected in 1930 when 135-ft. shaft sunk in search of gravel. (Logan 38:229.)
	Flora						See Buckeye Hill.
	Fort Jim		23	10N	11E	MD	Drift mine four miles southeast of Placerville. Active 1913-15.

PLACER GOLD (CONT.)

MAP NO.	CLAIM, MINE, OR GROUP	OWNER NAME, ADDRESS	SEC.	T.	R.	B & M	REMARKS
					LOCATION		
	Fossati (Tunnel)	Not determined	17	10N	12E	MD	Drift mine one and one-half miles south of Camino. Active intermittently 1930-36. Two channels, of Tertiary South Fork of American River, the lower being 25 to 200 ft. wide developed by adits and raises. (Logan 38:230.)
	Franklin (Tockey)	Placerville Gold Mining Company, Box 191, Placerville		10N	11E	MD	Drift mine two and one-half miles east of Placerville. Active around 1896 and in 1907. A Tertiary channel of the South Fork of the American River developed by 1400-ft. drift in channel. Gravel sent through 10-stamp mill and 100-ft. sluice. (Crawford 96:140-141; Tucker 19:301.)
	French Corral dredge	French Corral Gold Dredging Company, Sacramento					Operated dragline dredge at Brown's Bar on Middle Fork American River in 1946.
	General dredge	General Dredging Corporation, Natoma					Operated 1½ cu. yd., and 2 cu. yd. dragline dredges near Coloma and near Shingle Springs during 1939-42. Averill 46:255.)
	Gignac	George H. Wickes, 24 Olivene, Placerville	16	10N	11E	MD	Drift mine at Texas Hill two and one-half miles southeast of Placerville. Active during 1890's. (DeGroot 90:180; Tucker 19:301.)
	Giltedge	Not determined	9	8N	12E	MD	Two and one-half miles south of Fairplay. Active 1896. Tertiary channel capped with sand and clay contains coarse gold on bedrock. Developed by 300-ft. adit. (Crawford 96:142; Tucker 19:302.)

PLACER GOLD (CONT.)

MAP NO.	CLAIM, MINE, OR GROUP	OWNER NAME, ADDRESS	LOCATION SEC.	T.	R.	B & M	REMARKS
	Gold Bug (Canyon Creek)	Not determined	34	13N	10E	MD	Two miles north of Georgetown just north of Georgia Slide mine. Accumulated hydraulic mine tailings, seam deposit detritus, and some virgin gravel were intermittently worked by several concerns from around 1896 to 1934. Material was excavated by draglines and sent through trommel and sluices. (Crawford 94:105; 96:142-143; Tucker 19:301; Logan 22:301; 26:438-439.)
	Gold Channel Mining Company						Controlled drift and hydraulic mines in the Volcanoville-Kentucky Flat area that formerly were operated by the Two Channel Mining Company; which see. (Tucker 19:302.)
	Gray Eagle Bar						See Channel Bend.
	Gray Eagle Cliff			13N	11E	MD	At Volcanoville. Active about 1894. West-trending channel was worked through adit. Pay gravel is well cemented. (Crawford 94:112; 96:143; Tucker 19:302.)
	Great Bend dredge	Great Bend Corp., Lotus					Mined gold-bearing gravels at Lotus with gas shovel in 1931.
	Greenhorn dredge	Greenhorn Dredging Company, Auburn					Operated 2 cu. yd. dragline dredge on Middle Fork Cosumnes River near Youngs in 1940-42 and on Barkley property in 1947. (Averill 46:255.)
	Green Mountain	Not determined	17,20	10N	11E	MD	Drift mine on south side of Texas Hill two and one-half miles southeast of Placerville. The Green Mountain channel, a tributary of the Tertiary South Fork of the

PLACER GOLD (CONT.)

MAP NO.	CLAIM, MINE, OR GROUP	OWNER NAME, ADDRESS	LOCATION SEC.	T.	R.	B & M	REMARKS
	Green Mountain (continued)						American River, was mined through a 1700-ft. adit. (Lindgren 11:175-176.)
176	Grizzly Flat	Not determined	15	9N	13E	MD	Drift and hydraulic mine at Grizzly Flat. Active 1880's, 1896, and 1914-20. A north-trending channel on granite bedrock first worked by hydraulicking and later by drifting. Developed by 550-ft. drift. (Crawford 94:112-13; 96:144; Tucker 19:302; Logan 26:439.)
	Growers						See Claghorn.
	Harnish	Not determined		8N	12E	MD	Drift mine on Fairplay channel one and one-half miles south of Fairplay. Active 1896. Channel 4 to 5 ft. thick with much mica; developed by adit. (Crawford 96:144.)
177	Hayward (Indian Diggings)	Theresa Garibaldi, Amador City	18	8N	13E	MD	Drift and hydraulic mine at Indian Diggings. Active around 1896. A 275-ft. bank with 159 ft. of gravel; limestone bedrock. (Crawford 96:145, 147; Lindgren 11:181; Tucker 19:302.)
	Henness dredge	M. G. Henness and S. Legate, Georgetown					Operated dragline dredge near Georgetown during 1946-47.
	High Tunnel	Not determined	33	11N	11E	MD	Drift mine three miles northeast of Placerville and north of White Rock Canyon. Active in early days and 1926. Channel of Tertiary South Fork of the American River developed by 500-ft. adit. (Logan 26:439.)

PLACER GOLD (CONT.)

MAP NO.	CLAIM, MINE, OR GROUP	OWNER NAME, ADDRESS	SEC.	T.	R.	B & M	REMARKS
	Hinds (Los Angeles)	Alida L. Avansino, Route 3, Box 196, Placerville	29	10N	12E	MD	Drift mine one mile northwest of Pleasant Valley. Discovered 1927; active early 1930's. Bench gravel deposit lying on rhyolite developed by 48-ft. shaft and 200-ft. northwest drift. (Logan 38:233.)
	Hook and Ladder						See Toll House. (Averill 46:255-256.)
	Hoosier Gulch dredge	Hoosier Gulch Placers, Sacramento					Operated dragline dredge in Logtown Ravine in 1939 and near Shingle Springs in 1945 and 1947.
	Hope	Not determined		13N	11E	MD	Northeast of Volcanoville. A Tertiary channel deposit mined prior to 1892. (Preston 92:206; Crawford 94:113.)
	Hope						See Landecker.
	Horseshoe dredge	Horseshoe Dredging Company, Youngs					Operated dragline dredge near Youngs on Cosumnes River 1938-40. (Logan 38:363; Averill 46:256.)
	Horseshoe Flat	Not determined		10N	12E	MD	A drift mine two and one-half miles east of Newton; Long idle. (Tucker 19:302.)
	Hutchinson and Woodburn	Not determined	26	9N	12E	MD	A drift mine at Slug Gulch three miles northeast of Fairplay. Active 1926. (Logan 26:439.)
	Indian Diggings						See Hayward. (Lindgren 11:181.)
	Indian Diggings Creek	Not determined	18	8N	13E	MD	Hydraulic mine on Indian Creek near Indian Diggings. Active around 1896. Bedrock is limestone. (Crawford 96:147.)

PLACER GOLD (CONT.)

MAP NO.	CLAIM, MINE, OR GROUP	OWNER NAME, ADDRESS	SEC.	T.	R.	B & M	REMARKS
	Ingram dredge	W. E. Ingram, Gridley					Operated dragline dredge at Horseshoe Bar on Middle Fork of American River, 1940-42. (Averill 46:256.)
	Irish Creek	Irish Creek Mining Company, Georgetown					Operated non-floating washing plant near Georgetown in 1940. (Averill 46:256.)
	Irish Slide	Not determined	33	9N	13E	MD	Drift mine at Henry Diggings three miles south of Grizzly Flat. Intermittently worked since 1949 in conjunction with Payne and Christian drift mines by I. H. Campion, Somerset. See also Payne and Christian drift mines.
	Jerusalem	Not determined		10N	11E	MD	One and one-quarter miles east of Placerville. Prospected bench gravel deposit by hydraulicking in 1894. (Crawford 94:114; 96:147.)
	Jinkerson and Arditto	W. A. Jinkerson, 61 Del Monte Avenue, Los Altos	18	8N	13E	MD	Drift mine at Indian Diggings. Active 1913-17 and in 1926, developed by adit several hundred feet along. (Logan 26:439.)
178	Jones Hill	Not determined	20,29	13N	10E	MD	Hydraulic mine at Jones Hill five miles northwest of Georgetown. Active around 1892 and again in 1907. Channel trends northeast with slate bedrock; gravel 8 ft. thick. A 200-ft. bank hydraulicked. Equipped with 1200- and 1440-ft. flumes. (Preston 92:204; Lindgren 11:169.)

PLACER GOLD (CONT.)

MAP NO.	CLAIM, MINE, OR GROUP	OWNER NAME, ADDRESS	SEC.	T.	R.	B & M	REMARKS
	Kates (Norris)	Not determined		13N	11E	MD	Drift and hydraulic mine one and one-half miles east of Volcanoville. Active prior to 1894 and again in 1896 when prospected by Two Channel Mining Company. West-trending Tertiary channel deposit, 250-ft. bedrock adit and drifts. Gravel treated in stamp mill. (Crawford 94:114-115; 96:147,160.)
179	Kenna	Not determined	15	13N	11E	MD	Drift and hydraulic mine one mile northeast of Kentucky Flat. Active prior to and during 1896 when operated by Two Channel Mining Company. Some work done around 1922. Main or white channel hydraulicked; blue channel developed by 1500-ft. adit. Gravel well cemented and gold was coarse; gravel treated in 10-stamp mill. See also Two Channel Mining Company. (Crawford 96:160; Logan 38:251.)
180	Kentucky Flat	Not determined	22	13N	11E	MD	Drift and hydraulic mine at Kentucky Flat. Active 1894-1902 and again in 1933. Main or white channel hydraulicked in pit with 25-ft. bank; blue channel developed by 625-ft. adit and 80-ft. shaft. See also Two Channel Mining Company. (Crawford 94:115; 96:147,150; Logan 38:251.)
	Knight dredge	Knight Placer Mining Company, Georgetown					Operated dragline dredge in Georgetown area in 1947 and 1948.
	Kumfa (Kum Fa)	Not determined	15	10N	11E	MD	Drift mine at Smith's Flat. Active 1911-13, 1928 and 1936. Developed by 631-ft. inclined shaft; see also Carpender. (Lindgren 11:173; Tucker 19:302.)

PLACER GOLD (CONT.)

| MAP NO. | CLAIM, MINE, OR GROUP | OWNER NAME, ADDRESS | LOCATION | | | | REMARKS |
			SEC.	T.	R.	B & M	
	Kum Fu						See Kumfa.
	Lenčecker (Hope)	Not determined	20,21	10N	11E	MD	Drift mine one and one-half miles southeast of Placerville. Active in early 1900's and in 1925 and 1935. (Lindgren 11:178.)
	Last Chance	Not determined	28	9N	13E	MD	At Henry's Diggings two miles northeast of Omo Ranch. Long idle.
	Last Chance						See Try Again.
	Lemroh dredge	Lemroh Mining Company, San Francisco					Operated dragline dredge during 1939-40. (Averill 46:256.)
	Levenson			8N	13E	MD	Hydraulic mine one mile southeast of Fairplay. Active around 1896 when 50-ft. bank was mined. Gravel treated in 120-ft. sluices. (Crawford 96:148.)
	Lincoln dredge	Lincoln Gold Dredging Company, Lincoln					Operated dragline dredge in 1937.
181	Linden	Not determined	16,17	10N	11E	MD	Drift mine in Cedar Ravine one and one-half miles southeast of Placerville. Active 1882-94 when $130,000 was produced from gravel yielding $3.25 per yard. Deep Blue Lead channel developed by 4000-ft. adit with numerous

PLACER GOLD (CONT.)

MAP NO.	CLAIM, MINE, OR GROUP	OWNER NAME, ADDRESS	SEC.	T.	R.	B & M	REMARKS
				LOCATION			
	Linden (continued)						drifts along benches and two shafts. Gravel treated in 10-stamp mill. (Irelan 88:196-197; DeGroot 90: 179; Crawford 94:115-116; 96:148; Lindgren 11:177-178; Tucker 19:302.)
	Little Big Hole		30	9N	13E	MD	On Middle Fork Cosumnes River five miles northeast of Fairplay. Active 1926 when river diverted and 250-ft. adit worked. (Logan 26:440.)
	Lord and Bishop dredge	Lord and Bishop, 1853 North Eldridge, North Sacramento					Operated 3 cu. yd. dragline dredge on Greenwood and Carson Creeks in 1949 and 1950.
	Los Angeles						See Hinds.
	Lotus Bar		18	11N	10E	MD	Gravel deposit on South Fork of the American River at Lotus was mined by several concerns during 1930's. Gravels were excavated by power shovel or bulldozer and sent through washing plants. (Logan 38:235; 237.)
182	Lyon	Not determined	9,16	10N	11E	MD	Drift mine one mile southeast of Smith's Flat and two miles east of Placerville. Active prior to 1900 with a total output of $1,400,000. Deep Blue Lead channel with a number of winding benches developed by 2 shafts and drifts. (Lindgren 11:172,177.)
	Maple Leaf (Blakeley)	Not determined	11 or 12	10N	11E	MD	Two miles west of Camino near Five Mile House. Active in 1880's. From 1932 to 1935, operated as relief project and gave employment to 100 men. About $20,000

PLACER GOLD (CONT.)

MAP NO.	CLAIM, MINE, OR GROUP	OWNER NAME, ADDRESS	SEC.	T.	R.	B & M	REMARKS
	Maple Leaf (continued)						produced during this period by hydraulicking and sluicing.
	Matherly dredge	E. B. Matherly, Lotus					Operated suction dredge on American River near Coloma from 1947-1952.
	McCall						See Channel Bend. (Crawford 94:116-117; 96:150.)
	McCoy and Butler dredge	J.W.S. Butler, Sacramento					Operated dry land dredge in 1941.
	McKim	Not determined		9N	13E	MD	Drift mine near Henry Diggings. Prospected in 1926 when 100-ft. adit and 20-ft. raise driven. (Logan 26:440.)
	McQueen and Downing dredge	McQueen and Downing, Weaverville					Operated dragline dredge on Carson Creek in 1940. (Averill 46:256.)
	Mead dredge	Mead Company, 208 Fremont Street, San Francisco					Operated dragline dredge at Cherokee Bar in 1949.
	Mississippi	Not determined		13N	11E	MD	Drift mine one and one-half miles east of Volcanoville. Active around 1894. Tertiary channel deposit developed by 240-ft. adit. (Crawford 94:117-118; 96:150.)
	Molkey	Not determined		8N	12E	MD	Hydraulic mine one mile southeast of Fairplay. A north-east Tertiary channel lies on granite bedrock; much clay present. (Crawford 96:150-151.)

PLACER GOLD (CONT.)

MAP NO.	CLAIM, MINE, OR GROUP	OWNER NAME, ADDRESS	LOCATION SEC.	LOCATION T.	LOCATION R.	LOCATION B & M	REMARKS
	Mooney	Rhyolite, Inc., c/o E. C. Connella, 111 Sutter Bldg., San Francisco 4	15	10N	12E	MD	Drift mine two miles northeast of Newtown. Active 1894–96. (Crawford 94:118; 96:151; Tucker 19:302.)
	Morgan						See Two Channel Mining Company.
	Mount Gregory	Not determined	9	13N	11E	MD	Hydraulic mine three miles east of Volcanoville. Active 1896 and 1912 when 20- to 25-ft. bank with 8 ft. of cemented gravel was mined. (Crawford 96:151; Tucker 19:302.)
	Murzo (Brass)	W. E. Colby et al., 2416 East Flora, Stockton	12	13N	10E	MD	Drift mine on Buckeye Point one mile west of Volcanoville. Active around 1894 when 150-ft. adit driven in slate bedrock. (Crawford 94:119; 96:151.)
	Negro Hill		32	11N	11E	MD	At Negro Hill three miles northeast of Placerville. Tertiary gravel deposit mined by drifting and hydraulicking many years ago. (Lindgren 11:177.)
	Nelson dredge	R. N. Nelson, Sacramento					Operated dragline dredge near Shingle Springs in 1946.
	Norris						See Kates.
	Novis		9	13N	11E	MD	See Two Channel Mining Company. (Logan 38:251.)
	Old Empire	Not determined	32	9N	13E	MD	At Henry's Diggings one and one-half miles north of Omo Ranch. Long idle.

PLACER GOLD (CONT.)

MAP NO.	CLAIM, MINE, OR GROUP	OWNER NAME, ADDRESS	LOCATION SEC.	T.	R.	B & M	REMARKS
	One Spot (Sailor Jack)	Not determined	17,18	10N	12E	MD	Drift mine one mile south of Camino. Active in early days when $40,000 produced and again 1934-38. Two channels of the Tertiary South Fork American River one above other developed by 500-ft. adit with drifts and raises. Gravel yielded up to $8 per yard. (Logan 38:242.)
	Orloma dredge	Orloma Company, Placerville					Operated 1¼ cu. yd. dragline and dry-land washing plant on Indian Creek during 1941-42. (Averill 46:256.)
183	Pacific Channel (Zimmerman)	Not determined	34	11N	13E	MD	Drift mine one-half mile west of Pacific House. Active from around 1915 to early 1920's. Tertiary channel of South Fork American River trends southwest on granite bedrock and capped by andesite. Developed by several adits, one 1000 ft. in length. Gravel treated in barrel gravel mill. (Tucker 19:303; Logan 20:428-429; 26:440.)
	Pacific Dredge	Pacific Dredging Co.					Operated floating bucket-line dredge with 7½ cu. ft. buckets on Middle Fork American River at Mammoth Bar, 1914-18. (Logan 18:29.)
	Patterson	Not determined	20	8N	13E	MD	Drift mine two miles southeast of Indian Diggings. Active around 1935.
184	Payne	Mable Cole et al., 28 Crescent Drive Placerville	28	9N	13E	MD	Drift mine at Henry Diggings three miles south of Grizzly Flat. Active 1894; intermittently worked since 1949 by I. H. Campion, Somerset, in conjunction with Irish Slide and Christian drift mines. Tertiary channel deposit with 1 to 3 ft. of gravel. Developed by adit. See also Irish Slide and Christian drift mines.

PLACER GOLD (CONT.)

MAP NO.	CLAIM, MINE, OR GROUP	OWNER NAME, ADDRESS	LOCATION				REMARKS
			SEC.	T.	R.	B & M	
	Payne (continued)						(Crawford 94:120; 96:153; Tucker 19:302.)
	Pilot Hill dredge	Pilot Hill Mining Company					Operated dry-land dredge in Shingle Springs-Rescue area in 1935-36.
	Pleasant Valley	Not determined		10N	12E	MD	Tertiary channel deposit near Pleasant Valley prospected in 1894. (Crawford 94:121.)
	Potts and Maginess	Not determined		10N	12E	MD	Drift mine three-fourths mile east of Newton, long idle. (Tucker 19:302.)
	Quartz Canyon	Not determined		13N	11E	MD	One mile south of Volcanoville in Quartz Canyon. Accumulated debris worked in a self-shooting reservoir during the 1890's. (Crawford 96:155.)
185	Rising Hope	Roy C. Bishop, Box 70, Placerville	15	10N	11E	MD	Drift mine at Texas Hill three miles southeast of Placerville. Active 1910-20 and in 1929. Channel of Tertiary South Fork American River trends north, is 2 to 7 ft. thick, and up to 700 ft. wide. Near junction of Newton and Smith's Flat channels. Developed by 3500- and 3000-ft. drifts. Equipped with 50-ton barrel mill. (Tucker 19:303; Logan 20:429; 26:440.)
	River Pine dredge	River Pine Mining Company, Ltd., 141 Battery Street, San Francisco					Operated dragline dredge near Nashville in 1941-42, near Plymouth in 1946 and near Diamond Springs in 1949-50. (Averill 46:256.)
	Rivera	Not determined	16	10N	11E	MD	Drift mine at Texas Hill two miles southeast of Placerville. Active prior to 1900 and around 1905. Channel of

PLACER GOLD (CONT.)

MAP NO.	CLAIM, MINE, OR GROUP	OWNER NAME, ADDRESS	SEC.	T.	R.	B & M	REMARKS
	Rivera (continued)						Tertiary South Fork American River developed by 900-ft. adit, raises, and drifts. (Lindgren 11:178; Tucker 19:302.)
	Rocky Bar	Daily Journal Co., c/o R.C. Hoeschlaub, Suite 920, 621 S. Hope Street, Los Angeles	25	9N	12E	MD	On Middle Fork Cosumnes River near Cosumnes Copper mine and limestone deposit. Active early 1920's when potholes in limestone were worked for placer gold with pumps and derricks. (Logan 26:440.)
	Rogers						See Benfeldt.
	Roundout (Danaher)	Not determined		10N	11E	MD	Two and one-half miles northwest of Smith's Flat. Prospected around 1919. A 400-ft. inclined shaft and 600-ft. drift driven east in attempt to find gravel channel. (Logan 20:429-430; 26:438.)
	Rubicon	Volcanoville Mining Company, Room 910, Crocker Bldg., San Francisco	15	13N	11E	MD	Drift and hydraulic mine two and one-half miles west of Volcanoville. Active 1880's and 1890's. First worked as hydraulic mine. Gravel deposit 4 ft. thick and 27 ft. wide developed by 110-ft. drift. (Crawford 96:156.)
	Sailor Jack						See One Spot.
	Sailor Slide	Not determined	3 34	12N 13N	10E 10E	MD MD	One mile north of Georgetown in Georgia Slide area. Active 1919-22.
	Santa Rosa	Not determined	16	13N	11E	MD	Drift mine on Hopkins Creek one mile east of Volcanoville. Active 1894-96. Channel trends southwest; developed by

PLACER GOLD (CONT.)

MAP NO.	CLAIM, MINE, OR GROUP	OWNER NAME, ADDRESS	SEC.	T.	R.	B & M	REMARKS
	Santa Rosa (continued)						718-ft. adit. (Crawford 94:122; 96:156.)
	Schreiber dredge	A. L. Schreiber, Los Angeles					Operated dragline dredge near Coloma in 1942.
	Seils dredge	Seils Brothers, Auburn					Operated dry-line dredge near Auburn in 1939.
	Slug Gulch	Clifford Smith, Somerset, and others	25	8N	12E	MD	Hydraulic mine at Slug Gulch three miles east of Fairplay. Active in early days, during the 1900's, and again about 1930. Bedrock is partly limestone. (Lindgren 11:181.)
	Snow	Not determined		10N	12E	MD	Hydraulic mine one and one-half miles northeast of Newton. Active around 1896 when cemented gravel was mined in two pits. (Crawford 96:156-157.)
	Solari Tunnel		20	10N	12E	MD	Portion of Ventura drift mine, which see; prospected in 1935 through 351-ft. adit in search of bench gravel. (Logan 38:250.)
	Spanish Bar dredge	Inter-American Enterprises, Ltd., Sacramento					Operated dragline dredge at Spanish Bar on Middle Fork American River during 1950-51.
	Starbuck	F. M. Starbuck, Rescue	16,17	10N	9E	MD	Two miles northwest of Rescue. Gravel lying on decomposed granite intermittently mined by dragline late 1930's and early 1940's. (Averill 46:256.)

PLACER GOLD (CONT.)

MAP NO.	CLAIM, MINE, OR GROUP	OWNER NAME, ADDRESS	LOCATION				REMARKS
			SEC.	T.	R.	B & M	
	Stewart	Not determined	20	10N	11E	MD	Drift mine two miles southeast of Placerville between Chili Ravine and Weber Creek. Active 1880's and early 1890's. (DeGroot 90:180; Crawford 94:123; 90:157; Tucker 19:303.)
186	Texas Hill	Not determined	16	10N	11E	MD	Drift mine at Texas Hill two miles east of Placerville. Active around 1894. Channel of Tertiary South Fork of American River trends northwest, pay gravel 1 to 3 ft. thick, and capped by 120 ft. of andesite. Developed by 1500-ft. drift, 75-ft. incline, and 157-ft. shaft; breasts up to 100 ft. in width. Gravel treated in 10-stamp mill and 100-ft. suice. (Crawford 94:124-125; 96:159; Lindgren 11:173.)
187	Tiedemann (Tiederman)	Not determined	27	13N	11E	MD	Drift and hydraulic mine one and one-half miles south of Kentucky Flat. Active prior to 1896 and from 1896 to 1902 when operated by Two Channel Mining Company; again active 1932-1934. Main or white channel hydraulicked while blue channel developed by two adits, one 100 ft. in length. See also Two Channel Mining Company. (Crawford 96:160; Logan 38:251; Averill 46:257.)
	Tiederman						See Tiedemann.
	Tipton Hill	Not determined	34 ?	13N 12N	11E 11E	MD MD	At Tipton Hill seven miles northeast of Georgetown. North-trending channel deposit of tributary of Tertiary Middle Fork of American River mined by drifting and hydraulicking many years ago. (Lindgren 11:168.)
	Tockey						See Franklin.

PLACER GOLD (CONT.)

MAP NO	CLAIM, MINE, OR GROUP	OWNER NAME, ADDRESS	LOCATION				REMARKS
			SEC.	T	R	B & M	
188	Toll House (Hook and Ladder)	Charles Fossatti, Placerville	10	10N	11E	MD	Drift mine at Smith's Flat. Active prior to 1890, during the 1890's, and from 1918 to 1932. Deep Blue Lead and Gray Lead channels developed by 152-ft. shaft, raises, and several thousand feet of drifts. (DeGroot 90:179; Crawford 96:159; Tucker 19:303; Logan 20:428; 23a:45; 26:439; Averill 46:255-256.)
	Try Again (Last Chance)	Not determined	15	10N	11E	MD	Drift mine three miles east of Placerville. Active around 1890. Channel of Tertiary South Fork of American River trends northwest; developed by 1500-ft. bedrock adit and 213-ft. shaft. (Crawford 96:159; Lindgren 11: 179; Tucker 19:303.)
	Tunnel						See Fossati.
	Twin Forks dredge	Twin Forks Dredging Company, 1831 P Street, Sacramento					Operated dragline dredge on North Fork Cosumnes River near Youngs in 1949.
	Two Channel Mining Company			8N	12E	MD	Operated drift and hydraulic mines in the Volcanoville-Kentucky Flat area until about 1908. These properties included the Amelia, Bittera, Kates or Morris, Kenna, Kentucky Flat, Morgan, Novis, Tiedemann, and Wilton claims. (Crawford 96:160; Logan 38:251; Averill 46: 257.)
	Uncle Sam	Not determined		8N	12E	MD	Drift mine two miles south of Fairplay. Active 1896. Two adits run through bedrock to Tertiary channel. (Crawford 96:160.)

PLACER GOLD (CONT.)

MAP NO.	CLAIM, MINE, OR GROUP	OWNER NAME, ADDRESS	SEC.	T.	R.	B & M	REMARKS
	Union	Not determined	3 33	10N 11N	11E 11E	MD MD	Drift mine two miles east of Placerville. Channel of Tertiary South Fork of American River trends southwest, 6 in. to 4 ft. thick, 400 ft. wide, and has 100-ft. ande-site capping. Developed by 285- and 412-ft. shafts. (Tucker 19:303.)
189	Unity	Not determined	3,4	10N	11E	MD	Drift mine at Wisconsin Flat, two miles northeast of Placerville. Active early 1890's. Deep Blue Lead channel courses southeast; up to 12 ft. thick, capped with benches 200 ft. wide. Developed by 1700-ft. drift and inclined shaft. Gravel treated in 10-stamp mill and 160-ft. sluice. (Crawford 94:125-126; 96:161.)
	Uno	Not determined	9	8N	12E	MD	Drift mine three miles south of Fairplay. Active prior to 1896. Channel developed by 350-ft. adit partly in granite, 50-ft. drift and raise. (Crawford 96:161.)
	Van Dyke dredge	Van Dyke, Modrell, and Warner, Ione					Operated 3/4 cu. yd. dragline dredge on Gordon property in 1941. (Averill 46:257.)
	Varozza dredge	C.H.M. Mining Co., Sacramento					Operated dragline dredge during 1946-47.
190	Ventura (includes Solari Tunnel)	Robert Patterson, Bedford Avenue, Placerville	20	10N	12E	MD	Drift mine on north side ridge between Webber Creek and Pleasant Valley south of Newton. Active 1930's and early 1940's. A 1300-ft. adit in volcanic ash driven south in attempt to reach channel. (Logan 38:250,254; Averill 46:257.)

PLACER GOLD (CONT.)

MAP NO	CLAIM, MINE, OR GROUP	OWNER NAME, ADDRESS	SEC.	T.	R.	B & M	REMARKS
	W.W.	Georgetown Lumber Co., Georgetown	27	13N	10E	MD	Drift mine at Cement Hill four miles north of Georgetown. Active 1894. Channel coursing southeast. Prospected by 400-ft. bedrock adit. (Crawford 94:126-127; 96:161.)
	Wabash Deep Channel	Not determined	26	13N	10E	MD	Drift mine three miles north of Georgetown. Active 1856-67, 1907, and 1920. Channel developed by 4 shafts and 350-ft. bedrock adit. (Logan 20:430.)
	White Rock	Not determined		9N	8E	MD	One mile west of Clarksville. Gravels in Carson Creek mined by dragline in 1925-26. (Logan 26:440.)
	White Rock diggings	Not determined	31,32	11N	11E	MD	At White Rock Canyon three miles northeast of Placerville. Deep Blue Lead channel, mined by hydraulicking many years ago, yielded $5,000,000. (Lindgren 11:172, 176-177.)
	Wilton	Not determined		13N	11E	MD	Drift mine near Otter Creek eight miles northeast of Georgetown. Active around 1894. South-trending Tertiary channel 600 ft. wide developed by 700-ft. adit. See also Two Channel Mining Company. (Crawford 94:126; 96:160,161.)
	Worthington	Not determined		13N	11E	MD	Drift mine northeast of Volcanoville. Active around 1892 when gravel yielding 16 cents per cubic yard was mined. (Preston 92:206; Crawford 94:126.)
	Wulff	W. C. Wulff, Rescue	8	10N	9E	MD	Five miles northwest of Rescue. Active 1938-46. Surface placer worked by small-scale methods. (Logan 38:255; Averill 46:257.)

PLACER GOLD (CONT.)

MAP NO.	CLAIM, MINE, OR GROUP	OWNER NAME, ADDRESS	LOCATION				REMARKS
			SEC.	T.	R.	B & M	
	Yellowjacket	Not determined	18	8N	13E	MD	Drift mine near Indian Diggings. Active 1926-27 and 1930.
	Zimmerman						See Pacific Channel. (Tucker 19:303.)

SEAM GOLD

MAP NO.	CLAIM, MINE, OR GROUP	OWNER NAME, ADDRESS	LOCATION SEC.	T.	R.	B & M	REMARKS
	Beattie and Parsons Consolidated						See Georgia Slide. (Preston 92:203; Crawford 94:102; 96:133; Tucker 19: 281.)
	Blasdel	Not determined	35	13N	10E	MD	At Dark Canyon 2 miles north of Georgetown. Seam belt 2000 feet long trends N. (Logan 34:44.)
	Blue Rock	Not determined	34	13N	10E	MD	See Georgia Slide. (Preston 92:203; Crawford 94:104; Tucker 19:282.)
	Bower	Not determined	7	12N	10E	MD	At Greenwood. Active prior to 1892 with total output of $2,000,000. A 30- to 100-ft. zone of slate and schist contains quartz seams. (Preston 92:204; Crawford 96: 136; Tucker 19:282.)
	Carrol	Not determined	18	12N	10E	MD	On east slope of Greenwood Hill. Seam zone hydraulicked in open pit 40 ft. wide. (Logan 34:45.)
191	Castile	Not determined	27	12N	10E	MD	One mile east of Garden Valley. Two quartz veins in seam zone 18 ft. wide hydraulicked in early days. (Logan 34:45.)
192	Cedarberg (Drury)	Byron W. Bacchi et al., Lotus	1 36	12N 13N	9E 9E	MD MD	On east side American Canyon 2 miles northeast of Greenwood. First worked in 1868; active in 1890's and early 1920's. Veins and veinlets in slate trend N. and dip E.; yielded much specimen gold. Hydraulicked at first, later mined through a 318-foot inclined shaft with levels at 100, 200, and 300 feet. (Crawford 94:106; 96:137;

SEAM GOLD (CONT.)

MAP NO.	CLAIM, MINE, OR GROUP	OWNER NAME, ADDRESS	SEC.	T.	R.	B & M	REMARKS
	Cedarberg (Drury) (continued)						Tucker 19:282; Logan 26:413; 34:45.)
	Cranes Gulch (Whitesides)	Not determined	15	12N	10E	MD	One mile south of Georgetown. Active prior to 1874 when $100,000 produced from open pit 250 ft. long, 150 ft. wide, and 70 ft. deep. (Logan 34:45.)
	Drury						See Cedarberg.
	Fairweather and Fairweather No. Extension						See Spanish Group.
	Fisk						See Joseph Skinner.
	French (Nagler)	Marcus Peterson et al., Box 331, Suisun	13	12N	9E	MD	Just west of Greenwood. Active prior to 1874 and again during 1890's; total output of more than $500,000. Zone of quartz seams trends N. up to 200 ft. wide. Worked by hydraulicking in pit 80 ft. deep and 600 ft. long. Also developed by shallow shaft. (Preston 92:204; Crawford 94:110; 96:141; Tucker 19:286; Logan 34:35.)
193	French Hill	Harriett Taylor, 1612 E. 7th Street, Long Beach 4	36	13N	9E	MD	One mile southwest of Spanish Dry Diggings. Active around 1894. Quartz seams and 1½-ft. vein trend NW and dip NE. Mined in open cuts and material treated in 800-ft. sluice and 10-stamp mill. Also developed by 100-ft. shaft and 100-ft. adit. Also prospected for asbestos.

SEAM GOLD (CONT.)

MAP NO.	CLAIM, MINE, OR GROUP	OWNER NAME, ADDRESS	SEC.	T.	R.	B & M	REMARKS
	French Hill (continued)						See Asbestos section. (Crawford 94:110; 96;141; Tucker 19:286; Logan 26:404; 34:47; 38:207.)
194	Georgia Slide (Beattie and Parsons Consolidated, Blue Rock, Mulvey Point, Pacific)	Kelsey Lumber Company, 585 Main Street, Placerville	3 34	12N 13N	10E 10E	MD MD	(Tucker 19:286; Logan 24:8; 26:437; Knopf 29:48-49; Logan 34:36; herein.)
	Gilt Edge (Revenge Consolidated)	Not determined	20	12N	10E	MD	One mile southeast of Greenwood. Worked by hydraulicking in early days. (Logan 38:231.)
195	Golden State	Mamie E. Forni, Georgetown	29	13N	10E	MD	Just west of Jones Hill 4 miles northwest of Georgetown. Active early 1890's. Quartz seams in belt 200 to 300 ft. wide mined in open cut and sluiced. (Preston 92:204; Crawford 94:111-112; 96:143; Tucker 19:287.)
	Grit						Upper portion of deposit worked as seam deposit. See Lode gold section.
196	Hart	Daniel Gellerman, 4524 Brand Way, Sacramento	27	12N	10E	MD	(Logan 34:47; 38:233; herein.)
	Hines-Gilbert						See in Lode gold section.

SEAM GOLD (CONT.)

MAP NO.	CLAIM, MINE, OR GROUP	OWNER NAME, ADDRESS	LOCATION				REMARKS
			SEC.	T.	R.	B & M	
	Hodge and Lemon	Not determined	6	10N	11E	MD	At north end of Quartz Hill one mile north of Placerville. Seam zone mined in open pit 40 ft deep. (Logan 34:47.)
	Joseph Skinner (Fisk, Porphyry)	Not determined	7	10N	11E	MD	On Mother Lode one mile north of Placerville. Active 1896-98, 1901-03 and around 1932. Total output nearly $100,000. Ore in thin quartz seams and small bunches in dike rock and Mariposa slate. Originally worked by hydraulicking. Developed by 232-ft. adit and drifts. (Crawford 96:140; Tucker 19:285-286; Logan 34:28,45.)
	Lincoln	Not determined	34	13N	10E	MD	One mile northwest of Georgetown. Active 1896 and again 1926. A seam belt 300 ft. wide strikes north. Developed by 100-ft. cut and 3 adits, 150, 60, 110 ft. in length; 900 ft. of sluices employed. (Crawford 96:148-149; Tucker 19:289; Logan 26:414.)
	Little Chief	William Barklage et al., 5713 State Avenue, Sacramento	26,35	13N	10E	MD	On Canyon Creek two miles north of Georgetown. Active 1894. Seams developed by 130- and 240-ft. adits; ore treated in 1-stamp mill. (Crawford 94:116.)
197	Mameluke (Mameluke Hill)	L. L. Clark, 6 Arbor Way, Lafayette	3	12N	10E	MD	One mile north of Georgetown. Active prior to 1880 with total output of not less than $2,000,000. Coarse gold recovered from alluvium and seams in slate. (Whitney 80:118; Preston 92:203; Tucker 19:290.)
	Mameluke Hill						See Mameluke.

SEAM GOLD (CONT.)

MAP NO.	CLAIM, MINE, OR GROUP	OWNER NAME, ADDRESS	LOCATION				REMARKS
			SEC.	T	R	B & M	
	Morning Star group	Mary A. Fitzgerald, Box 35, Ivanhoe, California	2	12N	10E	MD	In seam belt 1 mile northeast of Georgetown. Active around 1926. Yielded about $75,000. (Logan 26:415.)
	Mulvy Point		35	13N	10E	MD	See also Georgia Slide. (Tucker 19:292)
	Nagler						See French.
	New Garibaldi	Henry Henze, 4571-A Chateau, St. Louis, Missouri	12	12N	9E	MD	One mile west of Greenwood. Active 1896. An 8-ft. deeply weathered vein in slate strikes northwest and dips northeast; ore shoot about 400 ft. long. Developed by 170-ft. crosscut adit and shallow shafts. Ore treated in 25-ton Kinkaid mill. (Crawford 96:152; Tucker 19: 293.)
	Pacific		34	13N	10E	MD	See also Georgia Slide. (Crawford 96:153; Tucker 19: 292.)
	Parsons						See Georgia Slide.
	Porphyry						See Joseph Skinner.
	Revenge Consolidated						See Gilt Edge.
	St. Lawrence	Not determined					Yielded $23,000 by hydraulicking. (Logan 34:47.)

SEAM GOLD (CONT.)

| MAP NO. | CLAIM, MINE, OR GROUP | OWNER NAME, ADDRESS | LOCATION | | | | REMARKS |
			SEC.	T.	R.	B & M	
198	Sam Martin	Claude A. Harris, 1011 Broadway, Vallejo	6,7	12N	10E	MD	One mile north of Greenwood. Active 1894-96. A 20-ft. zone of quartz seams in slate strikes northwest and dips 60° northeast. Upper portion sluiced, lower portion mined through 80-ft. adit. Was equipped with 200-ft. flume. (Crawford 94:122; 96:156.)
	Sliger						See Lode gold section.
	Snow Flake						See Spanish group.
199	Spanish Group (Snow Flake, Fairweather and Fairweather No. Extension)	E. F. Pease, 400 N. Michigan Avenue, Chicago, Illinois	12	12N	9E	MD	One mile northwest of Greenwood. Mined in early days: prospected in 1930's. Seam belt 100-ft. wide hydraulicked, yielded $13,600. Developed by drift adit. (Logan 34:47.)
	Swift and Bennett	Not determined	11	12N	10E	MD	Just south of Georgetown. Active 1870's, when rich ore produced. (Logan 34:47.)
	Waun	State of California	24	13N	9E	MD	One mile northwest of Spanish Dry Diggings. A seam belt 50 ft. wide mined by hydraulicking and later by underground methods. (Logan 34:47.)
	Whitesides						See Cranes Gulch.

IRON

MAP NO.	CLAIM, MINE, OR GROUP	OWNER NAME, ADDRESS	SEC.	T.	R.	B & M	REMARKS
200	Chaix	Roy Chaix et al., Box 87, Placerville	14	8N	9E	MD	One and one-half miles south of Latrobe. A lens of magnetite and hematite as much as 25 ft. wide crops out for a distance of 60 ft. Much silica present. Similar prospect 1 mile west. (Logan 26:441.)
	Gutenberger	Not determined					Five miles east of Diamond Springs. A prospect containing hematite. (Logan 26:441.)
201	Reliance	E. G. Wendt, Box 792, Rio Vista	18	10N	9E	MD	Three miles north of Bass Lake. Two magnetite veins 4½ ft. wide are in coarse-grained gabbro, which also contains disseminated magnetite. Developed by 50-, 218-, and 312-ft. shafts, now caved. (Aubury 06:297; Boalich 22:110; Logan 26:441.)
	Simons	Roy Chaix et al., Box 87, Placerville	13	8N	9E	MD	Adjoins Chaix prospect. A prospect containing magnetite and hematite. (Logan 26:441.)

LEAD

MAP NO.	CLAIM, MINE, OR GROUP	OWNER NAME, ADDRESS	SEC.	T.	R.	B & M	REMARKS
	Hazel Creek						A gold mine. Ore contains considerable amount of galena. See Lode gold section.

MANGANESE

MAP NO.	CLAIM, MINE, OR GROUP	OWNER NAME, ADDRESS	SEC.	T.	R.	B & M	REMARKS
202	Alderson	Albert Earl Harris et al., 69 Clay Street, Placerville	16 17	10N	11E	MD	One and one-half miles southeast of Placerville. Zone of arkosic sandstone 150 ft. long and cemented by black manganese oxide interbedded with Tertiary volcanic beds. Assays as high as 25 percent manganese. (Trask and others 43:111; 50:51.)
	Buffalo Hill	Not determined	10	12N	10E	MD	Just west of Georgetown. A prospect; manganese oxide in metasediments. Assays 11.7 percent manganese. (Trask and others 43:111; 50:51.)
203	David	Not determined	10	12N	10E	MD	Just west of Georgetown. Parallel bands of pyrolusite and psilomelane in NE-striking SE-dipping beds of chert and quartzite. Averages 5 to 10 percent manganese. Developed by open cuts. (Trask and others 43:111; 50:51-52.)
	Double E	Cecilia Simpson, 4029 Kuhule Street, Oakland	12	9N	10E	MD	Two miles southeast of El Dorado. Small low-grade deposit. (Trask and others 43:111; 50:52.)
204	Martinez	Martinez Gold Mines Company, P. O. Box 61, El Dorado	13	9N	10E	MD	Three miles southeast of El Dorado near Martinez gold mine. Two NE-striking lenses 2 ft. wide contain rhodonite and black manganese oxide. Country rock chert and schist. (Trask and others 43:111; 50:52.)
205	Moccettini	Robert D. Domecq, et al., Route 1, Box 39, Shingle Springs	13	8N	9E	MD	Two and one-half miles east of Latrobe. Greenstone stained with black manganese and iron oxides, developed by open cut. (Trask and others 43:111; 50:52-53.)

QUICKSILVER

MAP NO.	CLAIM, MINE, OR GROUP	OWNER NAME, ADDRESS	LOCATION				REMARKS
			SEC.	T.	R.	B & M	
	Amador						See Bernard.
206	Bernard (Amador)	J. L. Schneider, Sloughhouse	4	8N	10E	MD	By Fanny Creek 2 miles west of Nashville. Active during 1860's, some quicksilver ore produced; prospected 1903, again in 1917. Cinnabar with pyrite occurs interstitially in slate and quartzite. Developed by 75-ft. vertical shaft and 117-ft. adit. (Crawford 94:359; Aubury 03:190; Bradley 18:42; Tucker 19:306; Logan 26:445.)
	Shingle Springs	Not determined					Traces of cinnabar found 5 miles south of Shingle Springs. (Logan 26:445.)

TUNGSTEN

MAP NO.	CLAIM, MINE, OR GROUP	OWNER NAME, ADDRESS	SEC.	T.	R.	B & M	REMARKS
					LOCATION		
	Balderson	Not determined	10	12N	11E	MD	Three miles south of Balderson Station near Rock Creek. Minor amounts of scheelite in fine-grained tactite.
	Bear Creek						See Comeback Consolidated.
207	Bucks Bar	Martin Williams, Somerset	12	9N	11E	MD	Two miles west of Buck's Bar by North Cosumnes River. Small amounts of scheelite in contact metamorphic rock.
	Caldor						See Grizzly Flat.
208	Comeback Consolidated (Bear Creek)	Ajax Consolidated Mines Company, c/o Bruce Thompson, First National Bank Building, Reno, Nevada	4,	11N	11E	MD	(Logan 38:365; Averill 43:318; herein.)
209	Grizzly Flat (Caldor, Pierce, Sciaroni)	North portion: Caldor Lumber Company, Diamond Springs. Central portion: Horace Pierce, Sacramento; leased by E. L. Wilbur, North Sacramento. South portion: Americo and Columbus Sciaroni, Grizzly Flat	32 5	10N 9N	13E 13E	MD MD	Herein.

TUNGSTEN (CONT.)

MAP NO.	CLAIM, MINE, OR GROUP	OWNER NAME, ADDRESS	LOCATION SEC.	LOCATION T.	LOCATION R.	LOCATION B & M	REMARKS
210	Pacific House	Not determined	22	11N	13E	MD	One mile west of Pacific House on north bank of American River. Scheelite in tactite. Developed by open cut.
	Pierce						See Grizzly Flat.
	Pioneer-Lilyama		3	11N	9E	MD	Copper mine. Minor amounts of scheelite observed in tactite. See under Copper. (Cox and others 48:45.)
	Rubicon	Not determined	18	13N	12E	MD	Just east of the junction of the Middle Fork American and Rubicon Rivers. Scheelite found both in El Dorado and Placer Counties.
	Sciaroni						See Grizzly Flat.
211	Williams	Blanche Williams, Placerville	1	9N	11E	MD	Herein.

ASBESTOS

| MAP NO. | CLAIM, MINE, OR GROUP | OWNER NAME, ADDRESS | LOCATION | | | | REMARKS |
			SEC.	T.	R.	B & M	
	Contraband (El Dorado Copper Mining Company.)	Not determined	24	12N	10E	MD	See also under Copper. (Aubury 06:262; Logan 26:404; 38:207; herein.)
	French Hill	Harriett Taylor, 1612 E. 7th Street Long Beach 4	36	13N	9E	MD	See also Seam gold section. (Logan 26:404; 38:207; herein.)

DIMENSION STONE

| MAP NO. | CLAIM, MINE, OR GROUP | OWNER NAME, ADDRESS | LOCATION | | | | REMARKS |
			SEC.	T.	R.	B & M	
212	Sierra Placerite	Sierra Placerite, Route 2, Box 252, Placerville	29	10N	12E	MD	Herein.

DOLOMITE

MAP NO.	CLAIM, MINE, OR GROUP	OWNER NAME, ADDRESS	LOCATION				REMARKS
			SEC.	T.	R.	B & M	
	Larkin mine	Bernard B. Ball, 122 Bedford Avenue, Placerville	29	10N	11E	MD	See also Lode gold section. (Logan 47:233.)

LIMESTONE

MAP NO.	CLAIM, MINE, OR GROUP	OWNER NAME, ADDRESS	SEC.	T.	R.	B & M	REMARKS
	Alabaster Cave						See Rattlesnake Bridge.
	Auburn Lime Products Company						See Rattlesnake Bridge in text.
213	Browns Bar	Ideal Cement Company, 310 Sansome Street, San Francisco	4	12N	9E	MD	Deposit of bluish-gray limestone near Browns Bar by Middle Fork of American River. Apparently never worked. (Clark 54:465.)
214	Buckeye Canyon	Ideal Cement Company, 310 Sansome Street, San Francisco	34	13N	9E	MD	Deposit of bluish-gray limestone in Buckeye Canyon by Middle Fork American River. Apparently never worked. (Clark 54:465.)
	California Rock and Gravel Company						See Cool-Cave Valley.
215	Cool-Cave Valley (California Rock and Gravel Co., Cowell-Cave Valley, Cowell Mountain Quarries)	South 2/3 owned by Henry Cowell Lime and Cement Company, 2 Market Street, San Francisco North 1/3 owned by Ideal Cement Company, 310 Sansome Street, San Francisco	6,7, 18	12N	9E	MD	(Whitney 65:281; Hanks 84:107; Crawford 94:391-392; Lindgren 94:3; Crawford 96:628. Aubury 02:17; 06:68; Tucker 19:304, 390-391; Logan 21:43,432; Your; 25b:13 10; Logan 26:442-443; 27:282; Laizure 27:208; 29: 251; Logan 38:277,280; 47:222,224-226; Bowen and Crippen 48:73,80; Clark 54:438-465; herein.)
	Cosumnes						See Slug Gulch.
	Cowell-Cave Valley						See Cool-Cave Valley.

LIMESTONE (CONT.)

MAP NO.	CLAIM, MINE, OR GROUP	OWNER NAME, ADDRESS	LOCATION				REMARKS
			SEC.	T.	R.	B & M	
216	Diamond Springs	Diamond Springs Lime Corp., Diamond Springs	27,28	10N	11E	MD	(Crawford 94:391; 96:628; Logan 26:443-444; 38:274-276; 47:226-230; herein.)
217	El Dorado	El Dorado Limestone Company, Shingle Springs	15	9N	9E	MD	(Young 25a:1001-1002; Logan 26:441-442; 38:276-277; 47:230-231; herein.)
218	Indian Diggings	Theresa Garibaldi, Amador City, et al.	17,18	8N	13E	MD	Extensive deposits crystallized limestone at Indian Diggings and one mile to SW. (Whitney 65:281; Logan 26:444; 38:273; 47:231; herein.)
219	Marble Valley (Schwalin)	Henry Cowell Lime and Cement Company, 2 Market Street, San Francisco	8,17	9N	9E	MD	(Whitney 65:281; Logan 24:442; 38:280; 47:231-232; herein.)
	Mountain Quarries						See Cool-Cave Valley.
220	Rattlesnake Bridge (Alabaster Cave, Rattlesnake Bar)	Semon Lime Company, Auburn	15	11N	8E	MD	(Aubury 06:65-68; Tucker 19:304; Logan 21:432; 24:8; 26:442-443; 38:273-274; 47:222, 223-224; Clark 54: 461; herein)
	Schwalin						See Marble Valley.
	Semon Lime Company						See Rattlesnake Bridge.

LIMESTONE (CONT.)

MAP NO.	CLAIM, MINE, OR GROUP	OWNER NAME, ADDRESS	LOCATION SEC.	LOCATION T.	LOCATION R.	LOCATION B & M	REMARKS
221	Slug Gulch (Cosumnes)	Clifford Smith, Somerset	24, 25, 26	9N	12E	MD	(Logan 47:231; herein.)
	Vertin Lime Company						See Rattlesnake Bridge in text.
222	West Cool-Cave Valley	Ideal Cement Company, 310 Sansome Street, San Francisco	12	12N	8E	MD	One mile west of Cool-Cave Valley limestone by Middle Fork American River. A northwest-trending lens 450 ft. long and 50 ft. wide of bluish-gray, high-calcium limestone in metasediments. Old stone lime kiln nearby. See also Cool-Cave Valley herein. (Clark 54:460.)

SILICA

| MAP NO. | CLAIM, MINE, OR GROUP | OWNER NAME, ADDRESS | LOCATION | | | | REMARKS |
			SEC.	T.	R.	B & M	
	Brandon	Not determined	36	9N	9E	MD	Near Brandon Corner. Massive quartz vein 10 to 35 ft. wide crops out about 250 ft. (Logan 26:445; Sampson and Tucker 31:437.)
	Coon Hollow	Leased by Harms Bros. Construction Company, Sacramento	18	10N	11E	MD	At Coon Hollow 1 mile south of Placerville. Quartz cobbles produced from hydraulic mine tailings prior to 1926. At present deposit used as source of aggregate material. (Logan 26:445; Sampson and Tucker 31:445.)
	Snow	Placerville Gold Mining Company, c/o L. F. S. Holland, Box 191, Placerville	32	11N	11E	MD	At White Rock Canyon 4 miles northeast of Placerville. Massive quartz vein 25 ft. wide exposed 600 feet along strike; mined for silica at one time. (Logan 26:445-446; Sampson and Tucker 31:437-438.)

SLATE

MAP NO.	CLAIM, MINE, OR GROUP	OWNER NAME, ADDRESS	SEC.	T.	R.	B & M	REMARKS
	Buck	Pacific Minerals Co., Ltd., 339-10th Street, Richmond	36	11N	10E	MD	Adjoins Chili Bar slate mine. Active in 1880's. (Irelan 88:200; Logan 26:448.)
223	California-Bangor Slate Company	c/o Breed & Robinson, Financial Center Building, Oakland	2,3, 10, 11,14	11N	10E	MD	One mile northwest of Kelsey. Active prior to 1915. (Tucker 19:306; Logan 26:448.)
	California Slate Quarry	Not determined	23,25	11N	10E	MD	Three miles north of Placerville on north side of American River. Active around 1889. Slate of poor quality because of presence of pyrite. (Irelan 89:283; Aubury 06:150; Tucker 19:306; Logan 26:448)
	Chadbourne						See El Dorado Slate Products Company.
224	Chili Bar	Pacific Minerals Co., Ltd., 339-10th Street, Richmond	36	11N	10E	MD	(Irelan 88:199-200; Crawford 94:401-402; Aubury 06:150; Tucker 19:306; Logan 26:448; 38:365; Bowen and Crippen 48:68; herein.)
	El Dorado Big Tunnel Company						Near Chili Bar. Some development work around 1894. (Crawford 94:401.)
	El Dorado Slate Products Company (Chadbourne)	Not determined	6	10N	11E	MD	On south side Big Canyon 1½ miles north of Placerville. Roofing slate produced from several quarries during 1920's. Finished material sent across canyon via overhead cable. Waste sold for roofing granules. (Logan 26:449.)

SLATE (CONT.)

MAP NO.	CLAIM, MINE, OR GROUP	OWNER NAME, ADDRESS	LOCATION				REMARKS
			SEC.	T.	R.	B & M	
	Eureka Slate Quarry (Sierra Slate Corp.)	Not determined		11N	10E	MD	One mile south of Kelsey. Active from around 1886 to about 1926. Large amounts of dimension slate mined from quarry with 200-ft. face and up to 200 ft. deep. Slate delivered to Placerville via 13,000-ft. aerial tramway. (Aubury 06:150-152, 153; Tucker 19:306-308; Logan 22:602; 26:449-450; 38:365.)
	Landeker	Not determined		11N	10E	MD	Near Kelsey. Produced roofing slate during 1880's. (Irelan 88:199.)
225	Losh	Not determined	25	11N	10E	MD	One-half mile north of Chili Bar. Active 1890, 1921-24, and 1937. Dimension slate produced from open pit 50 ft. deep and 40 ft. wide. (Logan 23a:602; 26:450.)
	Pacific Minerals Company, Limited						See Chili Bar. (Logan 38:365.)
226	San Francisco Slate Company	Not determined	26	11N	10E	MD	On north side American River opposite Chili Bar. Roofing slate produced during 1890's from several quarries. (Crawford 94:401.)
	Sierra Slate Corporation						See Eureka Slate Quarry.

SOAPSTONE

MAP NO.	CLAIM, MINE, OR GROUP	OWNER NAME, ADDRESS	LOCATION SEC.	T.	R.	B & M	REMARKS
227	Bernett	D. P. Bernett, Shingle Springs	23	9N	9E	MD	Herein.
	Brandon (Richardson)	Not determined		9N	9E	MD	Near Brandon Corner. Active in 1920; 1 carload produced. A 2-ft. lens of gray soapstone developed by short crosscut adit and drift. (Logan 20:432; 26:451.)
228	Bryant	Bryant Ranch, Shingle Springs	9,10	9N	9E	MD	One-half mile north of Latrobe. Small amounts of soapstone mined by Industrial Minerals and Chemical Company, Berkeley, in 1954 for use as insecticide carrier.
	Darlington	Not determined		10N	11E	MD	Three miles southeast of Placerville near Webber Creek. Active in 1880's, sawed slabs produced. Lens of massive soapstone up to 25 ft. wide and 130 ft. long developed by open cuts. (Logan 26:451.)
	Harold	Not determined					An undeveloped prospect 4 miles from Shingle Springs. (Logan 26:451.)
229	Hayden	Mark Hayden, Shingle Springs; was leased by Industrial Minerals and Chemical Company, Berkeley	7	9N	10E	MD	Herein.
230	Pacific Minerals (Swift)	Rufus Swift, Shingle Springs; leased by Pacific	35	9N	9E	MD	(Logan 20:433; Logan 26:451; herein.)

SOAPSTONE (CONT.)

| MAP NO. | CLAIM, MINE, OR GROUP | OWNER NAME, ADDRESS | LOCATION | | | | REMARKS |
			SEC.	T.	R.	B & M	
	Pacific Minerals (continued)	Minerals Company, Ltd., 337 - 10th Street, Richmond					
	Richardson						See Brandon.
	Rossi						See Shingle Springs.
	Shingle Springs (Rossi)	Not determined		9N	10E	MD	One mile east of Shingle Springs. Active 1919-28 when soapstone shipped for use as roofing coating. Lens of green soapstone in serpentine developed by open pit. (Logan 20:432-433; 26:451.)
	Swift						See Pacific Minerals.

STONE

MAP NO.	CLAIM, MINE, OR GROUP	OWNER NAME, ADDRESS	LOCATION SEC.	LOCATION T.	LOCATION R.	LOCATION B & M	REMARKS
231	Anderson Pit	H. C. Anderson, Meyers	20	12N	18E	MD	One mile north of Meyers by U.S. Highway 50. Sand from decomposed granite quarried for use as road metal and aggregate in batching plant. Material excavated by bulldozer and loaded into dump trucks. Also, gravel produced from Truckee River near Meyers.
232	Butler Pit	Archibald Butler, Meyers	16	12N	18E	MD	Two miles north of Meyers by Truckee River. River gravel and sand from decomposed granite excavated for use as aggregate and road metal.
	Cold Springs Sand and Gravel Company	Jerry Brown, Placerville	9	10N	10E	MD	Herein.
	El Dorado County Road Department						Coarse sand mined for use as road metal and fill from several pits in decomposed granite in Deer Valley-Rescue-Shingle Springs area. It also is obtained from just south of Lotus and a few miles east of Aukum. Serpentine from Hummingbird Ranch near Garden Valley (see Hummingbird Ranch) and just north of Four Corners is used as road metal.
233	Harms Pit	Harms Brothers Construction Co., Sacramento	18	10N	11E	MD	Herein.
234	Highways, Division of	State of California	12	11N	10E	MD	Sand pit 1 mile northwest of Coloma by State Highway 49. Coarse sand derived from decomposed granite and granodiorite used as road fill and in asphalt mix. An extensive

STONE (CONT.)

| MAP NO. | CLAIM, MINE, OR GROUP | OWNER NAME, ADDRESS | LOCATION | | | | REMARKS |
			SEC.	T.	R.	B & M	
	Highways, Division of (continued)						quarry several acres in extent with 30-ft. faces is worked by power shovel; material is screened, and loaded into dump trucks.
235	Hummingbird Ranch	M. A. Nichols, Garden Valley	32	12N	10E	MD	Serpentine quarry 1 mile west of Garden Valley. Serpentine, used as road metal by El Dorado County Road Department, excavated by bulldozer and loaded into dump trucks. Open cut 200 ft. long, 30 ft. wide, and 10 ft. deep.